The Kingdom of Rarities

The Kingdom
of Rarities

Eric Dinerstein

Washington | Covelo | London

Library of Congress Cataloging-in-Publication Data

Dinerstein, Eric, 1952-
The kingdom of rarities / Eric Dinerstein.
p.   cm.
Includes bibliographical references and index.
ISBN 978-1-61091-195-5 (cloth : alk. paper)—ISBN 1-61091-195-4 (cloth : alk. paper)—ISBN 978-1-61091-196-2 (pbk. : alk. paper)—ISBN 1-61091-196-2 (pbk. : alk. paper) 1. Rare vertebrates. I. Title.
QL82.D56 2013
596—dc23
2012025535

Printed on recycled, acid-free paper ⊕

Manufactured in the United States of America
10 9 8 7 6 5 4 3 2 1

Keywords: Island Press, conservation, low-density population, top predator, trophic cascades, ecosystem engineer, rhino, tiger, jaguar, golden langur, saola, Kirtland's warbler, bird-of-paradise, New Guinea, rainforest, maned wolf, giant anteater, Bhutan, Nepal, rarity, abundance, speciation, endangered species, bushmeat, adaptive radiation, rarity, rare species, extinct species

To Roger and Vicki Sant, conservators of rarities

Look deep into nature, and then you will understand everything better.

—*Albert Einstein*

Contents

Acknowledgments

My immersion in the Kingdom of Rarities began when I was a twenty-two-year-old American Peace Corps volunteer appointed as survey ecologist of a newly created tiger sanctuary in the jungles of Nepal. Rare species have been my passion ever since.

In the intervening decades, I have had the privilege of relying on many experts as my generous and knowledgeable guides, most recently the individuals portrayed in this book. Mike Parr, Tom Brooks, and John Lamoreux taught me about rarity in birds and invited me to be a founding member of the Alliance for Zero Extinction. Bruce Beehler kindly shared his vast knowledge of New Guinea and has been a great birding companion. George Powell, Sue Palminteri, Gregory Asner, Robin Foster, and many colleagues from the Organization for Tropical Studies shared their insights into the rarities that populate rain forests. Sarah Rockwell, Carol Bocetti, and Joe Wunderle kindly educated me about Kirtland's warblers, and conversations with Gordon Orians, Richard Cowling, David Wilcove, Bruce Beehler, Don Wilson, and Peter Raven helped clarify the links between rarity and habitat specialization. Nancy Kittle and Leslie Coolidge made it possible for me to go to Grayling, Michigan, and see the Kirtland's warbler.

The rhino- and tiger-wallahs Andrew Laurie, Chris Wemmer, Mel Sunquist, Dave Smith, Hemanta Mishra, Anup Joshi, Shant Raj Jnawali, Vishnu Bahadur Lama, Harka Man Lama, Man Bahadur Lama, the late Gagan Singh, Ram Kumar Aryal, and Bivash Pandav introduced me to the world of jungle rarities in Nepal and the greater Terai Arc Landscape shared by Nepal and India. Carly Vynne and Mason, her canine companion, taught me how to track Cerrado rarities, and Leandro Silveira and his wife, Anah

Tereza de Almeida Jácomo, were generous hosts in Emas National Park. Edson Endrigo shared his knowledge of Brazilian "specialty" birds with me. Kent Redford, David Wilcove, and John Morrison helped me to see Emas in a special light. Liba Pejchar, Thane Pratt, Jack Jeffrey, and Paul Banko educated me about Hawaiian birds. David Hulse, Nguyen Tran Vy, Barney Long, Gert Polet, Craig Bruce, Nick Cox, and John MacKinnon shared their knowledge of Indochinese rarities. Kinzang Namgay, Mincha Wangdi, Nawang Norbu, and Sherub were excellent advisers and guides during my Bhutan wanderings.

Early drafts of this book benefited from the advice of Jonathan Cobb. Susan Lumpkin, Nancy Sherman, and Holly Strand, excellent editors all, helped tremendously in sanding a rough-hewn manuscript. My editor at Island Press, Barbara Dean, helped to conceptualize the book and refine it, and the support I received from David Miller and Erin Johnson was vital and timely. Many scientists reviewed one or more chapters, and I thank especially Mike Parr, Robin Naidoo, Eric Wikramanayake, Jared Diamond, George Powell, Sue Palminteri, Sarah Rockwell, Joe Wunderle, Carol Bocetti, John Lehmkuhl, Kent Redford, Peter Vitousek, Stuart Pimm, Barney Long, Liba Pejchar, Thane Pratt, Bruce Beehler, David Wilcove, John Seidensticker, Nick Cox, David Hulse, Jack Jeffrey, David Steadman, and Carly Vynne. Curt Freese read the entire manuscript and made many helpful suggestions. Pat Harris made many excellent suggestions to bring rarity into the clear.

I am deeply indebted to two scientists who have for decades been the leaders in thinking about rarity in nature, Kevin Gaston and Gordon Orians, for reading all the chapters and offering astute comments and keen insights. Gordon in particular made several passes through the manuscript, enlightening me along the way. Careers are shaped by great mentors, fine colleagues, and a healthy dose of good luck, and I have experienced all three. Finally, Jonathan Cobb helped me make sense of it all.

My good luck continued when I ran into Trudy Nicholson while we were both walking our dogs along Cabin John Creek, Maryland. Her illustrations capture a whimsy and poetry about rare wildlife that fills me with joy in what I do for a living. Chris Robinson gave geographic coherence to my wanderings with his maps.

My colleagues at World Wildlife Fund-US have been extremely supportive of my efforts to finish this book, especially Donna Kutchma—whose assistance was indispensible—George Powell, Eric Wikramanayake, and Robin Naidoo. This book is dedicated to Roger and Vicki Sant, members of the board of WWF-US, who have been so supportive of my efforts to save wild nature. Their generosity on behalf of saving life on Earth is extraordinary, matched only by their passion.

My mother, Eleanor Dinerstein, and my sister, Holly Dinerstein, have cheered me on over the years. My dogs, Ursie and Grace, have pointed out rarities on our walks and sometimes chased a few. Finally, my lovely, brilliant wife, Ute Moeller, put up with my quests to see rare birds, mammals, and plants and the painfully long time it took for me to write about them.

Chapter 1

The Uncommon
Menagerie

RIDING ON AN ELEPHANT'S BACK offers a privileged, if distorted, perspective on the natural world. Wildlife species that seem large and scary at eye level, such as rhinos and tigers, appear as miniaturized versions from this elevated vantage. My well-trained mount, Kirti Kali, plowed boldly through the dense twenty-foot-tall grasslands of Chitwan National Park in lowland Nepal, scattering spotted deer and wild boars in our path. On elephant-back one feels invincible. As we emerged from the tall grass into an open area, my driver, Gyan Bahadur, calmly steered Kirti alongside a rare greater one-horned rhinoceros—a dangerous species that, locally, tramples and kills several villagers a year. The two-ton female and her young calf continued grazing peacefully on the floodplain. The rhinos seemed oblivious to our presence because we had spent months habituating these aggressive creatures to close contact. As long as we remained on the elephant, rather than approaching on foot, the mother rhino would remain unfazed and we would stay in one piece.

My jungle wanderings also warped my perspective on how uncommon these animals had become. By 1988, at the end of my initial five years of research, I had recorded thousands of observations of Chitwan's 370 one-horned rhinos and had photographed, identified, and named nearly every one. Yet seeing them every day made me forget their global rarity. At the time, only about 1,500 survived in the wild worldwide; all but those in Chitwan roamed in Kaziranga National Park in northeastern India. Although the one-horned rhino's numbers have slowly increased since then—in 2012 there were over 2,900, distributed among twelve populations—this species remains among the most endangered large mammals on Earth. Its status during my initial study raised several questions: Had the ancestors of this rhino, a diverse ancient lineage, always been rare during their evolutionary history? Or is the rarity of the one-horned rhino a relatively recent phenomenon, triggered by habitat loss and poaching for the mythical qualities of the rhino's horn?

My ecological study took an unexpected turn when I asked Gyan to maneuver Kirti Kali into an ideal spot for a photograph. I raised my camera to capture an exquisite panorama: the rhino cow and calf in the foreground, perfectly framed by the Annapurna range and Mount Dhaulagiri to the north. Then I noticed some clumps of low trees that spoiled the picture's composition. Copses of a species called *bhellur* (*Trewia nudiflora*) stood out like archipelagoes in the midst of the grassland. I asked Gyan why the trees had assumed this pattern. He took a break from smoking a cigarette rolled in a jungle-leaf wrapper to answer my silly question. "Oh, it's the work of *gaida*," he said matter-of-factly, using the Nepali word for "rhino" and gesturing toward the tree islands. "Those are old rhino latrines."

Rhinoceroses return to the same places time after time to deposit their dung—not out of tidiness but because these communal latrines allow solitary animals living in dense vegetation to exchange vital data, via scents within the dung, about their whereabouts and sexual activity. The sheer size of the dung piles, sometimes dozens of meters long, and the dense stands of *Trewia* trees that sprang from them were a revelation to me. All the more so because when I first arrived in Chitwan, I had wondered how this giant herbivore could have even a minor influence in cropping the lush vegetation—the wall of green grass surrounding me—which was recharged each year by the summer monsoon.

The answer lay buried in the dung. By voraciously consuming *Trewia* fruit and defecating intact seeds in latrines scattered throughout the floodplain, the rhinos could rapidly convert the world's tallest grasslands into *Trewia* forests. Countervailing the rhino-dispersal effect were the annual floods, which wash away and bury *Trewia* seedlings, and the annual natural fires, which incinerate much of the previous year's crop. But some of these seedlings obviously survived to become tree islands. What remained as an indelible imprint for me was the staggering potential of rhinos to reshape their surroundings, implying, in this case, that ecological impact does not always reflect numerical abundance.

It would be a stretch to say that sifting through rhino dung or musing while on elephant-back triggered my fascination with rarity. But my observations of these rhinos, and observations that I and others had recorded of another globally rare denizen of their neighborhood, the tiger, made me wonder: What if more biologists fanned out to study in depth not the common mongoose or the ubiquitous spotted deer but members of Chitwan's uncommon menagerie—great hornbills, Gangetic dolphins, gharial crocodiles, sloth bears, and Indian bison? How might one's perspective on the natural world change? What novelties, complexities, and even counterintuitive elements might emerge, and what adventures lay in store for the pursuer of these rarities?

As a scientist, I knew that the interplay of rarity and abundance is central to understanding patterns of nature as well as understanding the idea of dynamic ecological balance. What do we mean by "rare," though? By what measure is a rhino or tiger considered rare? Most biologists would apply the term to a species that occupies a narrow geographic range, has a low abundance, or exhibits both traits. Often this label stems from a comparison of an uncommon creature with others that share its habitat or taxonomic group, but it can also be viewed in absolute terms. For example, sticking with rhinos, the greater one-horned rhinoceros is rare from a global perspective, with fewer than 3,000 individuals, but it's relatively common in comparison with the highly endangered Javan rhinoceros, of which fewer than 50 remain, and those restricted to one locale. In this book, I draw mainly on examples of rarity among mammals, birds, and plants—the creatures I know best. But the condition of rarity transcends appearance and taxonomy. Whether an organism has a backbone, a beak, pincers, or petals or is covered by scales, fur, feathers, or fins, the same rules apply—occupying a limited space geographically and exhibiting low population densities guarantees a place in what I call the Kingdom of Rarities.

The simple truth is that many, many species on Earth are rare, but few people other than biologists are even aware of this fact. A leading ecologist on the subject, Kevin Gaston, suggested an astonishing asymmetry of life on Earth: as few as 25 percent of the world's species, such as robins, rats, and roaches, may account for 90 to 95 percent of all individuals on Earth. But if Gaston's estimates are correct, as much as 75 percent of all species on Earth may be drawn from the ranks of the rare. It's a stunning idea to contemplate.

If relatively so few individual organisms on Earth make up the rare, why should biologists study rarity, the rhinos rather than the roaches? The obvious academic response is "Because we know so little about them." Rephrasing the question, though, brings into focus a profound and central riddle of nature: Why, wherever you land, do you always find a few superabundant species and a multitude of rare ones?

One of the first lessons in community ecology—the science of how species interact in nature—is the prevalence of rarity at any locale in the tropics. Sweep a forest plot with a butterfly net, identify all the trees in that tract, scan those trees for singing birds, and you'll find the same result: many individuals of a few species and a lengthy list of singletons. This pattern holds from the forests of Madre de Dios, Peru, to Mondulkiri Province, Cambodia. Even though rare species occur everywhere, we still know too little about how they fit into the big picture of our wild menagerie. But some intriguing answers have emerged regarding, for example, the roles various rare species play in shaping the form and functioning of ecosystems and how ecosystems are affected as particular rare species are lost.

Attention to rarity can raise vital questions: Are all rare species, for example, by definition on the verge of extinction? Have all species that are currently rare been historically rare? Which species common now are likely to become rare? Greater clarity on these fundamental issues will help shape our response to saving wild nature. Will species that are common now become rare as a result of changing climate? For example, how will egg-laying sea turtles find nesting sites when sea levels rise, and how will moisture-dependent frogs lay eggs when rain forests face prolonged droughts, in some cases by the middle of this century? When the microclimate at the summit of Mount Udzungwa in southern Tanzania changes in a profound way, will the African violet—ancestor of the familiar houseplant—and the Udzungwa partridge disappear, or will they be able to adapt to the new conditions?

During the 1980s, leading biologists began to suggest that we were in the midst of the sixth great extinction event in the history of Earth. And in 1995, Stuart Pimm, one of the fathers of modern conservation biology, calculated that the current rate of species extinctions was as much as 1,000 times the normal background extinction rate. If so, newly rare species may face different, and more serious, problems from those encountered by species that have historically been rare—another major reason for exploring rarity in the natural world.

Beyond the extinction crisis, some scientists refer to our current epoch, the Holocene, as the Anthropocene or the Homogenocene, terms that describe two aspects of a new ecological state that is still poorly understood. The first refers to our period, wherein the human footprint extends everywhere in nature. The second refers to another kind of affront in which certain species have spread or been introduced by humans far beyond these species' original range and, as a result, natural habitats around the world, full of invasive species, begin to resemble one another. Being rare in this brave new homogenized world, as we'll see in the case of Hawaii, could mean something much different from when these same species first appeared in relative isolation. Rarity is not just a condition of nature; it is a condition that can be—and has been—imposed on species by human activity, all too often sending them on the road toward endangerment and extinction. In short, viewing the natural world through the lens of rarity can bring certain facts and species traits to our attention that we might otherwise overlook. Understanding these facts and traits may in turn provide insights that can help us save species from the current state of environmental deterioration.

Many conservation biologists target "saving rare species" as the ultimate aim of their work. Yet rarity, as a phenomenon in nature, can take many forms, not only among species, although that is central, but also in the building blocks of the natural world: genes, populations of species, habitats, assemblages, and ecological and evolutionary phenomena. Species, with few exceptions, are made up of populations distributed across the landscape. Saving only one population of each rare species simply as a token gesture would be of little ecological value, especially where those species play a role in maintaining a given ecosystem's integrity. So an essential goal is to conserve multiple populations of species and the genetic, ecological, and behavioral features that these building blocks contain. Conserving dispersed populations and their genetic variability gives species a better chance of adapting to and persisting amid changing conditions, such as a rapidly changing climate or invasion of their homeland by introduced species.

Buried within the species extinction crisis is another, less publicized calamity: the increasing rarity of species populations. These losses of populations, as well as in some cases entire species, have led biologists to sound warning after warning. The eminent biologist E. O. Wilson, for example, pronounced in a speech in early 2000 that "biodiversity cannot afford another century like the last one. We are about to lose thousands of species a year, especially in rainforests." Wilson could have extended the depth of the problem, if risking the simplicity of his message, by adding a phrase whose meaning has gone unnoticed by the general public: we have been losing *populations* of species faster than we have been losing species themselves.

These two concerns—rarity of species and paucity of particular populations—merge when it comes to those species whose entire earthly existence is represented by a single population, as a result of either natural forces or human encroachment. Who are these singleton species, and how many of them are now close to the abyss of extinction?

In 2003, several colleagues and I put together a paper for the *Proceedings of the National Academy of Sciences* to address this question, name those species, and suggest how their imminent extinction might be prevented. Our work on the paper, which was published in 2005, sparked the scientific basis for this book, an interpretation of the evolutionary and contemporary aspects of rarity. We focused our effort on a subgroup of relatively well known but threatened vertebrates, our fellow creatures with backbones—birds, mammals, reptiles, amphibians (fishes are yet to be analyzed). We postulated that certain of these species were already so uncommon that they would be extinction's next dodo birds unless action were taken to prevent their disappearance.

To begin, we turned to the gold standard for evaluating rarity of wild species, the International Union for Conservation of Nature and Natural Resources (IUCN) and its famed Red List of Threatened Species, which ranks species on the basis of sizes of remaining populations. The IUCN assigns the category "endangered" or "critically endangered" to species whose numbers have plummeted toward extinction. We then went a step further. "Let's name the rarest of the rare, those species whose entire global range is limited to one population at a single site," my colleague John Lamoreux suggested. He was proposing that we limit our survey to such species as the Bloody Bay poison frog, which hails from the last patches of rain forest on the island of Trinidad, the only place on Earth where it can be found.

Once a species such as the Bloody Bay poison frog is restricted to a single dot on the map, if one or another of several catastrophes strikes—if the spot is plowed, burned, flooded, drained, paved, polluted, or overrun with pigs, rats, or other invasive species—the threatened species that lived in that dot is gone: vanished forever. Rarity then becomes the precursor to extinction or, at least, its preexisting condition. Alternatively, if you save the place, you save the rare species—conservation in black-and-white.

Our results provided some new insights and a number of surprises. First, despite there being 20,000 species on the IUCN Red List, only 800 species found at 600 sites (some species shared the same site) met our criteria. Second, half of the species limited to a single site turned out to be amphibians. Third, many single-population species were restricted to isolated mountaintops. A botanist on our team, George Schatz, cautioned the vertebrate specialists against any euphoric notion that saving the world's rarities might be as easy as saving some isolated mountaintops where few people live. "Remember," he warned, "the 250,000 or so vascular plant species have yet to be evaluated for levels of threat. At least 10 percent of these are known only from the single site where they were first collected." There is a joke among field biologists that rarity is partly a natural phenomenon and partly the result of some less energetic biologists failing to wander far enough from the road or the field station in surveying their specialty. There may be an ounce of truth to that, but the idea that the populations of many plant species, and the insect species they host, could be so few only reaffirms the important role of rarity, especially in the tropics.

The next question for our group of biologists was which rare species or place we should try to save first. This exercise drew us to a global map and triggered much debate. "Here." Mike Parr leaned over northern South America to point out the location of a mother lode of rarities. His pen tip lingered on a massif that stood by itself in northern Colombia, the Sierra Nevada de Santa Marta. The solitary giant sat about 42 kilometers from the Caribbean coast and about 115 kilometers from where the sawtooth eruptions of the northern Andean chain began. Santa Marta in Colombia, like Mounts Kilimanjaro and Udzungwa in Tanzania, Mount Cameroon on the border of Nigeria and Cameroon, and Mount Kinabalu in Sabah, Malaysia, are but a few of the dozens of solitary mountains in the tropical belt that are hotbeds of natural rarities. Why this might be so was one of the questions I wanted to investigate.

"Here is where I want to go next," I said, pointing to the Zapata Swamp on the island of Cuba. Considered the Cuban version of the Everglades, this freshwater swamp is home to the Cuban crocodile, the Zapata wren, the Zapata rail, and two species of hutia (a guinea pig–like rodent) found nowhere else in Cuba, the Caribbean, or anywhere else. In the same swamp are the only robust populations of several Cuban birds—the Cuban sparrow, Fernandina's flicker, Gundlach's hawk, and the blue-headed quail-dove—proving that rarity is not confined to tropical mountains or even rain forests.

As we populated the map in front of us and delved into the causes of rarity for the 800 species that met our conditions, we saw another insight into rarity confirmed. Some of these species had likely always been rare, such as the 13 frog species sharing the same genus and the same mountaintop in Haiti, the Massif de la Hotte; others on the list had been made rare by human activities. Some species had been much more common during an era when the climate was different from what it was during our mapping project—colder, hotter, drier, wetter. They were now climate refugees. Some species had been doing fine at a single site until rats arrived on their island. We realized that we had to consider all the different causes of rarity to better understand which species would be likely to persist without much conservation effort. We needed to know which species had always been rare but were now facing even lower numbers, a more limited range, or a new invader.

Some of the more promising places to look for the causes of rarity and of patterns of rarity and abundance are where there are no people. A remote mountainous region of New Guinea with no history of human visitation, the locale of chapter 2, offers a good venue to investigate the extent of rarity under natural conditions. By comparing what we discover there with what is found in other ranges where local tribes have access, we can begin to answer several fundamental questions about how rarity is created and what pattern exists where humans have had no perceivable influence. New Guinea also offers a rewarding glimpse of how extreme isolation and active geology can lead to rarity and a narrow range of resident species. In contrast, another area with low human activity, the Peruvian region of Madre de Dios, the locale of chapter 3, illustrates a condition that exists for many tropical rarities, from jaguars to canopy trees—a wide range of species living at extremely low densities.

The string of insults to nature brought about by human activities covers a staggering range including habitat loss, poaching and the consumption of body parts of rare creatures, introduction of diseases and invasive predators, expansion of agriculture to feed a growing human population, and the horrors of war. In this book I examine these human-induced causes of rarity, along with many natural influences, in a journey that spans most continents. In the natural world, the causes of rarity are often difficult to pin down or isolate to a single source. To untangle these strands, in each chapter that follows I sample different manifestations of rarity and consider probable causes and consequences for species and the ecosystems they inhabit. Much can be done in the short term to preserve species populations. Ultimately, though, the future of many species depends on our ability to live in greater harmony with the rare creatures among us. In Bhutan, the setting of chapter 9, where Tibetan Buddhism is the dominant religion and cultural conservation is part of the fabric of society, we see how rare species can persist and recover when humans coexist peacefully with wildlife and treat rare species with respect and compassion.

What is in store for rare species? Looking backward and examining evolution's fingerprints may provide some clues. The renowned ecologist Gordon Orians has noted that natural selection, as an evolutionary process, lacks foresight. It can't look ahead to help a species best adapt to a threat to its future survival, be it next year or several centuries or millennia hence. Thus, all the current traits and behavioral responses we see in such species as the maned wolf, the giant anteater, the rhinoceroses, and the Kirtland's warbler—all protagonists in this story—were shaped in their predecessors' environments. Yet some of those traits, even if selected for other reasons, may enhance persistence when a species becomes rare or, if it has always been rare, faces even more dramatic threats to its survival. Phrased another way, at least some species that have always been rare may possess traits that will allow them to hang on in the face of changing circumstances. In each chapter I examine such traits to assess whether such a repertoire, however unintended, enhances adaptation to life in the Anthropocene.

If the search for rarity and an understanding of its origins holds evolutionary interest and conservation importance, it also has a strong allure of its own. The truth is as simple as it is universal: we are seduced by rarity and novelty. Scientists live with this affliction, shared with art collectors, car buffs, and wine connoisseurs, many of whom are willing to pay exorbitant prices to add the rarest of items to their collections. The allure of the rare is what motivates many of us to raise a pair of binoculars—from the birder who scans the backyard feeder in hope of seeing an off-course migrant to the ornithologist who finds the now rare green peafowl in a Vietnamese jungle. Perhaps our search for rarity among wild things is a holdover from distant ancestors who sought to expand their monotonous diets, find new healing herbs, or discover a more potent aphrodisiac. A rare object might even have served as a status symbol and increased mating success. Whether stimulated by curiosity or by our most intense cravings, we humans, it seems, long to seek out what is scarce and, therefore, precious.

In the nearly forty years I have been studying rarity, a recurring fringe benefit has been the chance to visit exotic places and meet fascinating people in the search for spectacular species. I first heard the term "quest species" from Bruce Beehler, a scientist featured in chapter 2 who explored the most remote mountain range of New Guinea in search of rare birds of paradise. "A quest species," he imparted, "is a rare species, for sure. But it is also a near-mystical creature, one that shadows your existence, one that you must see before you die." Although avian specialists are famous for their single-minded pursuit of one bird or more for their life lists, they are far from unique. Primatologists scan the thickets for their quest mouse lemur. Herpetologists work the bushes for their prized chameleon. Botanists slog through swamps to find an orchid previously unknown to science. Even parasitologists seek their quest tick, embedded perhaps in the nether folds of a wombat or Tasmanian devil.

The study of rarity is of vital importance today, but it also allows us to glory in the extraordinary activity and variety of the natural world. Staring at a habituated rhinoceros in Nepal or contributing to a desk study on rarity, for example, can never replace the thrill of a first sighting: a rare species you have waited your entire life to see on its own terms, in its own place. A quest species, if you will.

I was on my way to the Amazon lowlands of southeastern Peru when I had the chance to see a rarity up close that I had always dreamt about. Before dawn, flashlights in hand, my guides led me to a bird blind at the edge of Manú National Park, where we waited

*Three male Andean cocks-of-the-rock (*Rupicola peruvianus*) singing,
with a female in the background*

for the show to begin. Few rare species seek more attention than
the flamboyant Andean cock-of-the-rock we had come to see. The
male's molten-orange plumage virtually glows in the dark. His vocalizations—a series of hoots, growls, and chimp-like whimpers—
accompany a ritualized shake of an unusual cowlick and rump. The
bird's name, dare one ask, is a reference to its habit of nesting in
rock walls rather than some biological double entendre.

The male's extravagant appearance flares when several of them
gather in the dank, kaleidoscopic undergrowth. As the dawn light
filters through the tropical highland forest of Peru, colorful bach-
elors scramble to their singing perches on nearby tree branches.
Biologists describe the location of the courtship that ensues as a
lek, a place where males congregate to advertise their individual
greatness. One bird triggers an explosion of song and dance that

lasts for minutes. Just as suddenly, they all go mute. Perhaps the shadow of an eagle has passed overhead? Then the cacophony resumes in earnest. Soon a drab maroon bird slips into the center of the gathering, sparking a more intense bout of singing and feather shaking. The female has arrived.

By 6:45 a.m., the males had quieted down and dropped into the dense foliage. I left the bird blind with my guides and strolled down the dirt highway to the nearby lodge. It's hard to avoid descending into cliché after witnessing a lek display of any bird or mammal. For me, it was a lifelong yearning now sated, replaced by a sense of awe in how evolution and the essential mission to procreate can go to such lengths.

My group enjoyed a celebratory breakfast in the café of the Cock-of-the-Rock Lodge. Accessible cock-of-the-rock leks in nature, such as the one we visited, are rare and usually reached only after a long hike. Over a second cup of coffee, the conversation spun in a widening gyre of questions: What if the glorious Andean cock-of-the-rock, one of the most colorful birds on the planet, were as ubiquitous as the house sparrow? Would anyone bother to look at it? Or would its fate be like that of the blue jay, a stunner for visitors to the United States but a backyard fixture evoking yawns from the locals?

Back on the trail, we heard the males start up another chorus. Left to their own devices, most rare species, like this charismatic Andean bird, would persist for several million years. A logical conclusion, one that will be explored and challenged in this book, is that rare species have adapted to cope with life at low densities, in small areas, or in restricted habitats. Unfortunately, wild nature is no longer being left to its own devices, and many species face a tenuous future. Our own species, now shooting past 7 billion and far from rare, faces a different challenge: how to live sustainably without destroying the last strongholds of rarity. For rare species, the struggle is to hang on for dear life until, one day, humans gain the wisdom and humility to share nature's kingdom.

Chapter 2

The Gift of Isolation

IN 1704, A SCOTTISH BUCCANEER by the name of Alexander Selkirk was abandoned by his captain on a remote, uninhabited island off the coast of Chile. Selkirk had become a nuisance aboard the *Cinque Ports*, telling his fellow mates that the ship was unseaworthy. None heeded his warnings nor his invitation to join his party of one; all preferred instead to sail on with their commander. It was a bad choice in the end: the *Cinque Ports* later dashed on the rocks and many of her crew members drowned in the surf.

Soon after Selkirk headed inland to seek shelter, a large, reddish hummingbird flitted past him. How strange to encounter a hummingbird on an island in the South Pacific Ocean nearly 700 kilometers off the coast of Chile. Yet the Juan Fernández firecrown (named for the cluster of islands that included Selkirk's new home) was one of the few native vertebrates sharing this remote outpost with Selkirk and some feral goats introduced by earlier sailors. Selkirk spent the next four years as a castaway, alone, living off goat meat, wild fruit, and greens. He was eventually rescued by one of his former shipmates and returned to England to some acclaim.

The tale of the hummingbird and the marooned sailor took place on an island called Más a Tierra, which might seem irrelevant except that it now goes by the name of Robinson Crusoe Island. Literary historians have hypothesized that Daniel Defoe's novel, first published in 1719, was inspired by Selkirk's ordeal. Others have disputed this claim, pointing out that Defoe set his hero down on a tropical shore resembling Tobago or Trinidad rather than the more temperate Juan Fernández Islands. Regardless of the venue, *The Adventures of Robinson Crusoe* remains one of the most widely read pieces of fiction of all time.

The hummingbird has drawn less fanfare, but it has its own interesting side story. Unlike Selkirk, the firecrown was a longtime resident, having arrived on the island slightly less than a million years earlier. The firecrown is endemic to Robinson Crusoe Island, meaning that this island is the only site on Earth where it can be found. The total firecrown population is estimated to be 700 to 2,900 individuals, down from as many as 10,000 individuals earlier in the twentieth century and considered critically endangered today. Another, related hummingbird found on the island, the green-backed firecrown, arrived in the nineteenth century. Unlike its close relative, the green-backed is widespread in its range, being common across Argentina and Chile, and it is easily twice as numerous as the endemic hummingbird on Robinson Crusoe Island. Among the 330 species of hummingbirds in the world, though, these are a rare brace, the only two to have reached an oceanic island, reproduced, and gone on to live such a remote life.

Early nineteenth-century naturalists surely read Defoe's classic novel. And the most famous, Charles Darwin, passed near what is now Robinson Crusoe Island as the HMS *Beagle* set sail for the Galápagos Islands from coastal Chile. If Darwin had visited, it's hard to imagine his not mentioning with interest the presence of hummingbirds on an oceanic island, so far from the mainland. The modern explanation of why one hummingbird species is found only on a dot in the South Pacific while another, closely related species has a much broader range and is much more abundant is straightforward. The endemic red-backed species, the Juan Fernández firecrown, evolved over time into the distinct species it is on the islands, whereas the green-backed is too recent an arrival on the island for it to have diverged from its mainland relatives. For many other closely related species, however, the riddle of why some are common while others are rare remains to be answered.

Endemism epitomizes island life, especially along the equator. And among biologists, tropical islands crowd the atlas of daydreams—not just for the idyllic scenery but also because the unusual, and often rare, plants and animals inhabiting such islands showcase the arc of evolution. This scientific fascination has noteworthy milestones. Darwin's visit to the Galápagos and Alfred Russel Wallace's journey to the Malay Archipelago two decades later gave rise to these two men's vision of evolution by natural selection, the greatest organizing principle of life on Earth and one that sheds much light on the issue of rarity. Natural selection, at its essence, is the evolutionary process that results in living creatures becoming well suited, or adapted, to their locale. Darwin and Wallace's theory views evolution by natural selection as the dynamic force by which individuals that possess certain advantageous features, or traits, are more likely to survive to bestow these heritable traits, and duplicates of the genes that code for them, on future generations. The sum of these traits increases what is called the reproductive fitness of an individual, and the metric biologists use to assess this quality is the offspring one leaves behind and their ability to survive and breed.

Biologists still marvel over the explanatory power of the theory of evolution by natural selection, a lens through which to view every aspect of the natural world, from the habitats wild species select to the foods they eat, the mates they choose, the places they sleep, and their responses to predators. Yet few are aware that its coinventor Charles Darwin was also one of the first scientists to highlight rarity in nature. "Rarity," he stated in a neglected sentence from *On the Origin of Species* in 1859, "is the attribute of a vast number of species of all classes, in all countries." Had he expanded on this theme—perhaps in a sequel called *On the Origins and Ubiquity of Rare Species*—the topic of rarity might have become more central to the scientific orthodoxy much earlier.

Rarity is all around us, but an excellent starting point for understanding its generation, superior to the Juan Fernández group in its illustrative qualities, is on a large, highly mountainous tropical island. Even better is if such a landmass is far from a major continent and has remained separated from it for eons. In this environment we could seek a rich assemblage of plants and animals illustrating the first condition of rarity—a narrow range. Such an ideal locale sits on the other side of the southern Pacific Ocean from the Juan Fernández Islands: the huge island of New Guinea. Here we can clearly observe how geologic events and evolutionary mechanisms influence rarity; indeed, they help to create it. New Guinea's array of fabled birds of paradise and tree kangaroos illustrates how such episodes and processes lead to the separation of populations and the surfacing of new species, a number of which earn the moniker "rare."

Geologic events are harbingers of evolution. By this I mean that they often create the conditions that divide previously continuous populations into subpopulations that are isolated from one another, the condition that enables them to evolve separately. In New Guinea, for example, the formation of rugged mountain ranges isolated populations of the same species from one another by distances so great that individuals no longer dispersed between them. Over many generations, through genetic mutation and adaptation to differences in the environments on the isolated ranges, the populations may diverge enough in important characteristics that if members of the divergent groups are reunited, they can no longer interbreed. In this way new species are born—a process called speciation—with such new species often starting their existence as rare forms with a narrow range and low numbers. For this process to occur for birds and mammals, the isolated populations must persist long enough for the required genetic changes to accumulate. Speciation in birds and mammals does not happen on small islands because distances are not great enough to isolate populations for a long enough time. In contrast, amphibians do speciate on small islands, owing to their reduced mobility. But New Guinea is large enough and rugged enough for its lofty mountain ranges, surrounded by tropical lowlands unsuitable for mountain animals, to have witnessed many bouts of speciation. Although much of the seminal work on speciation used islands as the model, speciation is hardly restricted to islands and occurs on large landmasses as well. So, a critical insight about rarity is that island life per se is not the key to frequent evolution of new species and new rarities; rather, it is isolation, which can be provided by island archipelagoes or large islands.

If all islands are physically isolated from a mainland, are all islands, by the very nature of their limited range, repositories of rarities? Not necessarily, because several factors influence evolutionary processes on islands. Among them are distance from a mainland, length of time the island has been isolated, and size and diversity of habitats on the island. Islands that were once connected to the mainland are called continental islands. Some of them separated from a mainland when the ancient continent of Gondwanaland split up, beginning about 200 million years ago. These include Madagascar, New Zealand, the Seychelles, New Guinea, and New Caledonia, off the northeastern coast of Australia. They are loaded with ancient endemic species, many rare, and most quite different from the closest mainland flora and fauna. Other continental islands separated from adjacent mainland only a few thousand years ago when rising sea levels caused by melting glaciers severed their lowland connections. For example, Sri Lanka was connected to India only a few thousand years ago, and Trinidad was connected to

Venezuela until about 11,000–15,000 years ago. On each of these islands, the flora and fauna are quite similar to those of the continent nearby because insufficient time has elapsed for much evolutionary change to happen.

Other island groups, such as the Hawaiian and Juan Fernández Islands, are oceanic islands formed by volcanic action. They were never connected to a continental landmass. Thus, all organisms living on them must have dispersed across oceanic barriers. This is why the Juan Fernández Islands, although formed 1 to 6 million years ago, have few endemic vertebrates and only a modest number of endemic plants. Also, the few islands in the archipelago are too close to one another to isolate populations. The fauna and flora of nearshore oceanic islands, such as those in the Gulf of California, often tend to look quite similar to those on the mainland. They are within easy reach of mainland species that can fly or raft over on floating vegetation or whose seeds arrive on the winds. Habitat diversity also plays a key role in the amount of speciation that takes place. Like New Guinea, some of the most ecologically diverse islands—such as New Caledonia and Madagascar—are covered in mountain chains or bisected by plateaus. The resulting rain shadow creates wet forests on the island's windward side and dry forests on its leeward side, the different habitats favoring different species.

New Guinea not only embodies areas of exceptional isolation conducive to speciation. Much of the island has another quality that makes it an ideal natural laboratory for the study of rarity and abundance—remote mountain ranges marked by the virtual absence of human interference. Today, natural patterns of rarity and abundance on virtually all equatorial islands have become increasingly obscured by the destructive spread of invasive species—goats, cattle, pigs, cats, dogs, rats, rabbits—and by the logging of native forests, conversion of cutover land to agriculture, and other forms of development and exploitation that have followed the arrival of humans.

A scientist in New Guinea, in contrast, can still observe the interplay of geology, evolution, and rarity in an all but undisturbed venue. The mountain walls, deep gorges, and numerous rivers create barriers that prevent recently arrived species from spreading and swamping the local biota. Some of the high ranges even check the spread of humans. New Guinea is about twice the size of California but remains sparsely populated. With so few people—about 7 million inhabitants—and so much forest and rugged terrain, there may even be places where people have never set foot.

Among wild destinations, New Guinea surpasses all others as an outpost of mysterious dimensions. The lack of roads and few airstrips limit access to its isolated valleys. Flights into the mist-cloaked mountains are fraught with danger. In 1991, an up-and-coming field biologist and a colleague, Ian Craven, perished when his single-engine bush plane crashed in the island's far western wilderness. Then there is the famous disappearance of anthropologist Michael Rockefeller about fifty years ago somewhere on the southern coast. Explorers of New Guinea know the risks and the challenges: avoiding deadly strains of malaria; living on tinned mackerel and navy biscuits; not getting lost in the uncharted forest; and not getting eaten. The highlands are known for their fierce mountain clans who wage ritualized war with neighboring groups and occasionally dine on one another.

Then there are the rewards. On my first trip, in 1990, I carried along a copy of *Birds of New Guinea* by Bruce Beehler and Thane Pratt. The pages of the bird guide and Dale Zimmerman's illustrations came to life when I saw and heard my first magnificent riflebird, a bird of paradise, in a park outside the capital. Upon my return, I finally met Bruce, and discovery of our mutual interests in New Guinea's unique flora and fauna led to more frequent contact. I grew envious listening to his stories about what he had observed—the birds of paradise, cassowaries, bowerbirds, giant fruit bats, tree kangaroos—and his biological surveys into the most remote region of the planet.

~

The island of New Guinea is especially interesting to biologists because so many of its species are found nowhere else. New Guinea itself is politically divided—the western portion is Papua (formerly Irian Jaya), a province of Indonesia, and the eastern half is the sovereign nation of Papua New Guinea, or PNG for short. The political division obscures a common geography, similar rain forests, and shared cultures. No other large tropical island is so mountainous, and the isolation created by its cordilleras, or mountain chains, has had a profound effect not only on the evolution of animal and plant life but also on human communication. Nearly one-fourth of all spoken languages on Earth are known only in New Guinea (about 1,100 true languages, not including dialects), and most are spoken by fewer than 1,000 people. Languages, like birds of paradise or tree kangaroos, can also be labeled as endemic to a place. And perhaps the same forces—geologic, geographic, and evolutionary—that resulted in so many tongues spoken by so few people in this land might be related to why so many species of widely different lineages occupy such narrow ranges: a prime element of rarity.

In the late summer of 2005, I received an excited message from Bruce that he would have to bow out of a birding trip we had planned in Maryland. "I can't believe it," he wrote. "After twenty-five years of trying, I have just been granted permission to bring a small research team into the Foja Mountains!" More than twenty years of guerilla warfare in Papua Province had prevented any field expeditions in this region. When Papua's political troubles subsided in 2003, Bruce joined herpetologist Stephen Richards of the South Australian Museum to try once more. Two years later, in October 2005, villagers of the Kwerba and Papasena clans granted Bruce and his group permission to enter their homeland. It would turn out to yield a wonderful collection of naturally engendered rarities.

If New Guinea is the ultimate destination for field biologists, within it the Fojas loom as the pinnacles of desire. The Foja Mountains were reportedly so inaccessible that humans had never settled there. I had come to doubt whether places such as the Fojas still existed, geographic outliers with no history of interlopers—gold miners, oil drillers, religious zealots, or armed guerillas—either seeking their fortune or looking for an escape from modern society.

The purpose of Bruce's expedition was to survey the biota and to find species new to science and others poorly known that he thought might be inhabiting this isolated range. The Fojas sit in the heart of Papua Province, and their summits reach 2,200 meters above sea level. The surrounding 7,500 square kilometers, lightly inhabited by jungle dwellers, lack roads. Taken together, the vast landscape stands as the largest expanse of pristine forest in the tropical Pacific.

The Fojas have a reputation for repelling outsiders. The legendary secretary of the Smithsonian Institution S. Dillon Ripley, for whom Bruce had worked, tried to approach them from the north in 1960. He failed because the rivers were not navigable. In the late 1970s, both Bruce and Jared Diamond, a noted New Guinea bird expert long before penning his best-selling *Guns, Germs, and Steel*, raced to explore them. Diamond, through helicopter and grit, arrived first in 1979. He found a species that had eluded birders for seven decades, the "lost" golden-fronted bowerbird, and returned home to bask in ornithological glory. His find earned extensive press coverage, and his technical paper reporting the rediscovery of the bowerbird enlivened the cover article in *Science*.

Bruce believed that the Fojas might hold many species new to science as well as others that science had forgotten, the so-called lost species. Although they had not been declared extinct, these "lost" species had not been seen in decades, a category of rarity but a half step from oblivion. To help them, Bruce and Steve Richards assembled a team of crack naturalists who specialized in different taxa—birds, mammals, reptiles, amphibians, butterflies, other insects, plants—along with several Papuan biology students. They would be guided by members of the Kwerba and Papasena tribes who lived in the Foja region. The guides were as excited as the biologists at the prospect of exploring this area. Not only were the Fojas uninhabited, but also—as far as they knew—the mountains had never been part of their clans' hunting territories.

Would new species of apes or monkeys be spotted in the Fojas? Would the researchers need to be vigilant for prowling leopards or tigers? If this were Sumatra, such concerns would be accurate. Tigers, leopards, apes, and monkeys, however, are part of the Indo-Malayan but not the Australian faunal realm. New Guinea sits east of what is known as Wallace's Line; most of the islands west of that demarcation were physically connected to Asia (via a landmass known as the Sunda Shelf) during recent glacial periods when sea levels were much lower than they are today. Asia's large vertebrates had no need to swim across even shallow waters to reach Sumatra, Java, Borneo, and Bali. They simply walked.

Technically, Wallace's Line falls between the islands of Bali and Lombok and divides the Indonesian archipelago in two. Alfred Russel Wallace was the first to identify this natural longitude, so the name celebrates his insight. On the western side of Wallace's Line live animals of largely Asian origin; on the eastern side are those common to Australia, to which New Guinea was at times connected (to the Sahul Shelf) by a land bridge over the Torres Strait prior to about 20,000 years ago. The lack of land bridges spanning Wallace's Line and the presence of deep ocean trenches precluded the mixing of faunas. The simple fact is that most mammals are poor long-distance swimmers, and they can't drink seawater. So even those that managed to hitch rides on floating mats of vegetation—natural rafts—would have died of dehydration before they reached land. In New Guinea, then, birds of paradise and tree kangaroos would be evident, but no orangutans, gibbons, macaques, or leaf monkeys ever arrived there. The range of large terrestrial carnivores also stopped farther west, in Bali (tigers) and Borneo (clouded leopards), because they too could not cross deep water. Absent as well would be the giant herbivores, such as tapirs and wild cattle, that the big cats prey upon.

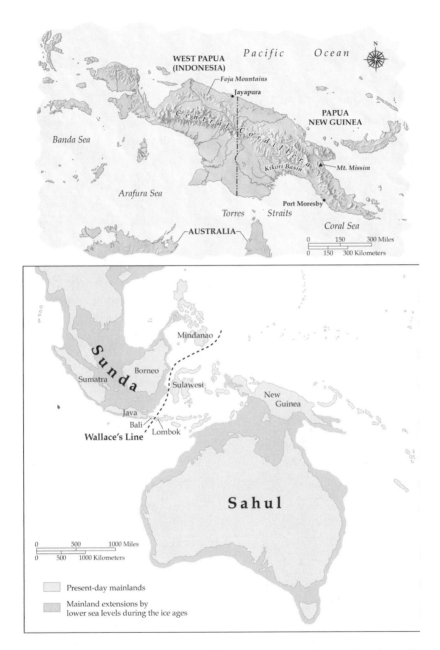

Top: map of the island of New Guinea and environs; bottom: locations of Wallace's Line, the Sunda Shelf, and the Sahul Shelf

~

Bruce's expedition to the Foja Mountains held out the promise of enabling us to assess the accuracy of how rarity in relation to abundance is commonly portrayed in natural tropical habitats—a few abundant species and lots of others represented by a few scattered individuals or small groups. Perhaps absent humans, especially local hunters, a different pattern would emerge from his team's observations in virtually undisturbed terrain.

Bruce's New Guinea team assembled in Jakarta, Indonesia's capital, in early November. On November 12 they flew from Java to Jayapura, the capital of Papua Province and the city nearest to the Fojas. During the ensuing three days of cutting through red tape with provincial bureaucrats, Bruce dealt with a new worry. Before arriving in Papua, he had been unaware that helicopter service was so scarce; transport might not be available to lift the team in. To come this far only to lose the sole mode of access to the summits would be a cruel blow.

As his colleagues grew restless, Bruce managed with great effort to secure a helicopter, only to encounter another restriction. The permits limited the expedition to a rapid assessment. The company could offer transport for only two specific days: *in* on November 22 and then *out* on December 7. The team would have to work fast and under pressure.

In the first stage of the expedition, they flew by small plane to a foothills airstrip in the Kwerba homeland. There they spent a week exploring the steep lower Fojas, collecting specimens, and planning the next leg. On November 21 they packed carefully and prepared for the helicopter to lift them higher up the mountains. Torrential rains marred the evening and increased the fear that their helicopter would be grounded on the following day. Bruce and Steve decided to split the overall expedition into two teams. The hill-forest team would take a short hike from Kwerba and explore the lower slopes. Bruce would lead the helicopter-mountain team. "No problem in splitting the group, as many do not want to go in helicopter into high mountains . . . (not safe)," he wrote in his field journal, the short parenthetical postscript capturing the fears that went unspoken.

When their transport arrived at 9:30 a.m. on November 22, a few team members immediately set off with Bruce and the pilot into a heavy mountain cloud. They looked out the window in vain for their intended "landing spot," a boggy lakebed. Not the ideal helipad, but a rare patch of flat earth amid the rugged mountains. In the fog, however, they couldn't see a thing. During twenty minutes of searching, the anxiety levels on the chopper climbed steadily. After a few more turns with no visibility, the pilot opted for an instrument landing and located the site using the global positioning system (GPS). The cloud opened just as he lowered to drop them in.

Some tree kangaroos lounging in the branches noted the noisy intrusion, if not its significance. The men now unloading boxes of camping gear represented the largest-ever group of scientists and possibly the first humans to walk on this spot, 1,650 meters above sea level. The most ecologically important reading was not the altitude, however; it was the distance from the nearest village: a two-week trek. Three additional runs, navigating the ever-increasing cloud on the mountaintop, brought in the rest of the helicopter-mountain team. The final run, carrying much of the botanical and mammalogical field equipment, did not make it in because of the weather. The team would have to do without.

To the newcomers, the absence of old campfires and forest trails—signs of hunters at work—was encouraging. The two headmen from the local Kwerba and Papasena tribes accompanying the expedition were as astounded as the biologists at how isolated the place was. As far as they knew, no one from either clan had ever been here. Perhaps birds of paradise, bowerbirds, tree kangaroos, and spiny anteaters—the species that Bruce and his colleagues hoped to see, on the basis of Bruce's previous work and Jared Diamond's reports—would still thrive here. These were the species that would offer clues about the patterns of rarity and abundance among closely related species, especially in the absence of humans.

Setting up camp in the forest by the bog dispelled some of the landing jitters, and any lingering fright soon gave way to elation. The first afternoon, as Bruce marked trail routes for animal surveys, a song from the forest suddenly stopped him in his tracks. He could barely contain his joy. He had just heard the call of the black sickle-bill, a bird not expected to be in the Fojas. One of the most sought-after birds of paradise—a glossy black creature with a sickle-shaped bill, red irises, and a saber-like tail—it is also the most difficult of its family to see in the wild.

The challenges of the landing and the potential for imminent new discoveries propelled everyone on the team to begin pursuing their specialties, assisted by graduate students from The State University of Papua (UNIPA). Steve Richards, the herpetologist, expected a welcoming chorus of new frog species. Kris Helgen was eager to see which species of tree kangaroos and other rare mammals awaited him. Wayne Takeuchi readied his plant presses for a bonanza of new species. Brother Henk van Mastrigt gathered his butterfly nets. Bruce focused on birds.

The next morning's first sighting bolstered Bruce's earlier prediction that the Fojas were probably high enough and expansive enough, and certainly isolated enough, to be a source of many new discoveries. Right in camp, a bird appeared that looked like a new species of honeyeater. The black, grackle-sized bird sported a face mask of extensive orange wattles. When the wattles became engorged with blood, the honeyeater looked as if it were blushing. The features failed to match any of the species Bruce and Thane Pratt had described or Dale Zimmerman had illustrated. This unique bird, later named the wattled smoky honeyeater, became the first new avian species discovered in New Guinea in fifty-four years.

The last described bird species in New Guinea had been the long-bearded honeyeater, discovered in 1951 by E. Thomas Gilliard on Mount Hagen, another isolated mountain peak. Among the vertebrate hunters, ornithologists have now been almost everywhere on the planet. The Fojas may be unique in being uninhabited and untrammeled, but they join many other tropical mountain ranges in supporting rare species restricted to a single mountaintop or range. Biological expeditions to other parts of New Guinea and to New Caledonia, the Andes, Mount Kinabalu, the Eastern Arc Mountains of Tanzania, and the Annamite Mountains of Vietnam have time and again discovered new species on tropical peaks.

To understand this phenomenon—how species evolved into new forms in mountainous regions—evolutionary biologists and rarity specialists need to summon their inner geologist. In the Annamites of Vietnam, in Hawaii, in any place of volcanic origin, or where Earth's plates have shifted, the causes of rarity in tropical landmasses can go back eons.

New Guinea's geologic history is one of great complexity, and the birth of the Foja range illustrates this point. It all began with tectonic plates, the unevenly shaped floating slabs of rock that sit under the continents and oceans. Tectonic plates trace back to the early formation of planet Earth, almost 5 billion years ago, and their motion has been compared to that of slow-moving bumper cars—colliding, separating, colliding again and remaining stuck together—with the movements causing the continents to drift. When they collide, new landmasses arise. Less than a million years ago, ongoing contact between the Australian and Pacific tectonic plates uplifted deep seafloor material. From the collision of the two plates, the Fojas were born and rose rapidly (by geologic standards) out of the ocean. This uplift became an isolated landmass and attracted mountain-dwelling species from adjacent mountain ranges. Over cycles of cool and warm climate, mobile species from the high cordillera to the south that runs the length of New Guinea were able to colonize the newly rising Foja Mountains and other isolated ranges. Once settled on this recently uplifted chain, these populations of plants and animals were slowly carried upward into the mists as the mountains inched higher year by year. In isolation from others of their species and through the process of evolution by natural selection, these populations evolved unique traits.

This uplift created the series of mountain islands where species evolved, resulting in changes in some aspect of their appearance from the first arrivals. The geologic account reinforces a near-immutable law of rarity: the isolation of tropical mountain ranges leads eventually to the creation of new species and, in turn, drives patterns of rarity. This is because—by definition—when one species diverges from another in an isolated habitat, it often starts out as rare in terms of its narrow range. In some cases, a collision of continental plates may cause uplift that separates one widespread, abundant species into two populations, which may diverge. In general, however, many species on tropical mountains lack the dispersal capability to rejoin their former populations. Even tropical birds that could fly over the mountains tend to avoid such flights if it means leaving the altitudinal belt in which they are most accustomed to live.

Bruce's field journal included a most-wanted list. Truth be told, the expedition was for him an intensely personal pursuit of a quest species, the golden-fronted bowerbird, endemic to the Fojas. It was a dream shared by many. This species had not even been described until 1895, when Lord Walter Rothschild identified it from trade skins collected by locals from somewhere in western New Guinea. The discovery inspired many naturalists to search for the bird, but over the following eighty-two years all returned from their expeditions without success. Consider this entry by Michael Everett in his book *The Birds of Paradise and Bowerbirds*, published more than three decades ago: "Nothing is known of the Golden-fronted Bowerbird, except for what can be learned from only four known museum skins, all males . . . it has never been found in the wild . . . and among bowerbirds or birds of paradise is unique in this respect. It may be either a very rare bird or one which is virtually extinct—but it is certainly one which has a very restricted range."

The bowerbirds must have been eavesdropping. The next day, the team found several display bowers of this elusive species close to camp. According to the *Oxford English Dictionary*, a bower is "a pleasant shady place under trees or climbing plants in a garden or wood." The bowers these birds construct are remarkable, fashioned from moss and sticks and decorated with blue and yellow fruit. The males spend hours each day vocalizing from perches in the saplings beside their bowers, often imitating the songs of other birds as well as other natural sounds. At the sight of the bowers, Bruce felt elated at the prospect that his quest might soon be realized. Now if they could just see the bird!

The next day, while marking survey trail routes through the forest with colored flagging tape, Bruce stumbled upon a male in attendance at a bower. The first encounter took his breath away. The robin-sized male sports a brilliant bonnet of yellow feathers, a look emulated by some New Guinea tribal chieftains who fashion the feathers into headdresses. When the yellow feathers are erected during its mating dance, the bowerbird is one of the most compelling sights in nature.

The oblivious male began emitting a rasping growl that sounded a bit like gravel being poured out of a dump truck. This vocalization helped Bruce and his team find many other bowers. The dancers themselves could be easily spotted circling around their one-meter-tall fabrications. In the tranquility of the Fojas, the bowerbirds proved anything but shy.

A day later, it was the turn of another quest species to make its grand entrance. A pair of Berlepsch's six-wired birds of paradise appeared at camp, cavorting in from the forest edge. Equipped with shiny black-and-white plumage, the male began to shake his feathers, including the six ornate wirelike feathers emanating from his head, the basis for the name of the species. He then began to flick his wings, flash his white flank plumes, and utter sweet call notes while closely circling the female. The display lasted for more than five minutes.

The scientists were awestruck. No one had witnessed the courtship dance of this species before. The six-wired's place of residence was unknown before this expedition; it was a species collected before 1897 from an uncertain locale and not by a naturalist. No one had even managed a good look at the adult male—Jared Diamond had caught only a glimpse of this "lost" bird, a female, in his earlier expedition. It had been listed as a subspecies of another six-wired for nearly a century, but Bruce instantly recognized it as a potentially distinct species because of its unique vocalizations and slightly varied plumage and eye color. Given so many isolated ranges on this island corrugated with mountain chains, it's no wonder that New Guinea is home to thirty-seven species of birds of paradise. Bruce and his team worked straight through the daily downpours. Over the next two days the marsh in the bog flooded, turning the helicopter landing site into a lake and forcing him to move his tent and much of the campsite to higher ground. On the plus side, the mystery bird with the chicken wattles, the new honeyeater, was everywhere around the newly flooded bog. Oddly, it made no vocalization, an unusual trait for a bird.

There was another striking aspect of the birds up here. Previously, Bruce had logged many hours of observation concealed in blinds on Mount Missim, about 750 kilometers away in PNG, recording the breeding and feeding behavior of four local species, both rare and common, of secretive birds of paradise. But in the upper Fojas, a bird blind was unnecessary. The male Berlepsch's six-wired bird of paradise ignored the visitors and continued to dance vigorously. The golden-fronted bowerbird males could be observed constructing their bowers up and down the ridgelines, and by the end of week one, the wattled smoky honeyeater had become a camp bird. Nearby, the black sicklebill continued flirting with the females from atop a dead stub on a forest ridge.

What made the birds so unafraid of predators, nonhuman or human? Bruce's field notes from the expedition indicated relatively few hawks, falcons, and eagles during his bird surveys. Perhaps their densities were too low to provide much of a threat to the displaying birds. Then, too, there were none of the midsized and smaller wild cats that frequently prey upon birds west of Wallace's Line. As for the human influence, one might expect much more secretive behavior if hunters regularly passed through. Finally, if such breeding displays are genetically hardwired, the birds can't help themselves from becoming momentarily oblivious to everything else. In the desire to breed with a female, what Darwin termed "Nature's urge," being a shy wallflower conferred no selective advantage. A male had to step out on the dance floor if he wanted to mate and pass on a copy of his genes to future generations.

Birds of paradise have fascinated indigenous peoples and scientists alike, and those on Bruce's team were no exception. The sixteenth-century Spanish explorers who named the species they encountered believed that these birds were emissaries from heaven. Some Europeans in the nineteenth century even assumed that birds of paradise were ethereal, legless creatures that never touched ground. Birds of paradise are native to New Guinea and surrounding islands, with only a few species resident in northern Australia. Many in this family of forty species dazzle biologists with their bright, metallic wirelike feathers, brilliant gorgets (throat patches), and elaborate tails. Their mating dances, full of shakes and shimmies, put those barbs and bristles in best light to advertise their fitness to interested females. Today most species are not threatened and some are widespread, while others fit the definition of rarity by occupying slivers of altitudinal ranges along the flanks of the steep mountains.

Back at camp, Bruce thumbed through his own field guide to explain the evolution of this group to the several Papuan students gathered round. There are good reasons why the story of these spectacular birds illustrates so well the links between evolution, narrow ranges, and rarity. *Birds of New Guinea* indicates the tremendous variation in the most obvious traits of plumage color and size, as well as in less prominent features such as beak length and size of feet.

Evolution is not sorcery, but the transformation of a single bird of paradise ancestor into these forty wondrous variations is magical just the same. Biologists, though, use a different term for "abracadabra"—

adaptive radiation. This is the evolutionary phenomenon by which a number of species evolve from a single ancestor, often diverging to occupy different ecological niches and isolated geographic spaces. Genetic evidence shows that birds of paradise probably descended from a crow-like bird about 28 million years ago when the bird of paradise line split off from the ancestors of the crows (the so-called corvine assemblage that arose in the Australian region). In the absence of primates and other bird groups as predators, birds of paradise spread their wings, so to speak. They flew to new places and then changed in certain traits, often in the presence of and in competition with other resident species. Once reproductive isolation from members of their species they had left behind had been established in their remote mountain domains, they evolved to expand their diets in response to food sources available in their new homes, whether as fig eater, or bark chipper and grub extractor, or spider hunter, or miner of specialized fruit hidden in woody capsules of mahoganies and nutmegs. Some forms developed stronger legs to walk along tree branches and pluck fruits and insects, for example, while others developed short, weak bills and specialized in soft fruits. To apply the founding principle of natural selection: those that diversified in bill size or shape, and as a result fed more efficiently in their new habitat, invariably left more surviving offspring, while other variants over time passed from the scene.

Although the honeyeaters of the South Pacific rarely receive mention in the textbooks, they are another classic example of adaptive radiation in birds, with about 184 species descended from a single ancestor. The wattled smoky honeyeater hanging about Bruce's camp represented one of the newest species in the family. As we saw earlier, some isolated species, such as the Juan Fernández firecrown, remain rare, while a few, such as the green-backed firecrown, become transplanted or jump the ecological barriers and become abundant and widespread. Such shifts in rarity and abundance depend on a host of natural and human factors explored in later chapters.

During his daily searches, Bruce had begun tallying the number of birds of paradise he spotted or heard singing in the dawn chorus. By the second week in the Fojas, he had observed in miniature the general pattern of distribution in New Guinea. Some species, such as the two manucodes, the lesser and king birds of paradise, the riflebird, and the pale-billed sicklebill, were confined to the lowlands and foothills of the Fojas. Other species, including the buff-tailed sicklebill and the several kinds of six-wired birds of paradise, restricted themselves to bands of forest at middle elevations. Still others, such as the black sicklebill that Bruce heard on the first day, holed up only on mountaintops. So it turned out that over time the birds of paradise had sorted by what biologists call altitudinal stratification. Perhaps some species were better than others at taking advantage of what nature gave them at different elevations and became further adapted to prospering in narrow elevational bands of forest. Above or below their particular elevational band, they might have been replaced by another species.

As a group, birds of paradise have no rivals—except perhaps the pheasants—and are the most beautiful birds in the world. There is widespread use of their plumage among New Guinea natives, who fashion headdresses of the feathers to wear at clan gatherings. Some clansmen prize the long enameled, pearly head feathers of the king of Saxony bird of paradise. Others go for the white, trailing plumes of the ribbon-tailed. Highly valued, too, are the iridescent blue breast feathers of the superb bird of paradise. Beauty and extreme rarity in New Guinea often coincide, with the result that some of these species have been wiped out in areas around villages. A Papuan ornithologist documented the carnage wreaked by such ornamentation, reckoning that about 36,000 birds had been killed to furnish headdresses for one clan gathering in one mountain town.

Each evening back at camp, the botanists also shared their discoveries. Tropical botanists often feel overshadowed on expeditions, perhaps because their quarry is stationary. Even if a spectacular species is in flower or fruit, the display is ephemeral, unlike the headdress of the bowerbird or the pelage of the golden-mantled tree kangaroo. But here in the Fojas, botanists on the team came upon an eye-catching plant that rivaled anything found by the vertebrate chasers. It was a rare white-flowered rhododendron and featured the largest bloom on record for that genus, measuring almost eighteen centimeters across. (Wild rhododendron flowers are seldom more than six centimeters in diameter.) The flower came from a canopy-living shrub-like rhododendron that also appears as a shrub around bogs. Rhododendrons are common at high elevations, yet another example of adaptive radiation, this time in a common genus of shrubs, trees, and climbers. In all, New Guinea is home to about 164 of the nearly 850 species of the genus *Rhododendron*; most are found in the uplands, as in such places as the Fojas, where three unique species can be found.

The alpine zone above timberline in New Guinea's cordillera, including the Foja Mountains, is quite small compared with those of the Himalayas and the Andes, so the alpine plants endemic to this habitat would have very limited ranges. Robert Johns, one of the fathers of botanical research in New Guinea, estimates that there may be as many as 15,000 to 20,000 species of plants endemic to New Guinea and its outer islands, or about 5 to 8 percent of all vascular plants on Earth. About 70 to 80 percent of New Guinea plant species that are endemic have very limited ranges, most in the alpine zone. In contrast, the canopy tree species in the lowland rain forests have wide ranges. The most common ones share an affinity for the mainland tropical forests of peninsular Malaysia and the remnant rain forests of northeastern Australia.

Plants offer yet another perspective on the link between endemism and rarity on islands or other geographically isolated locales. Consider the plants on some more familiar islands: Puerto Rico, Jamaica, Hispaniola, Cuba, the Galápagos, New Caledonia, New Zealand, Hawaii, and even the Juan Fernández. Among all these islands, the Hawaiian chain, not the fabled Galápagos, has the highest percentage of endemics. Hawaii has a flora of about 970 vascular plant species, with 91 percent found only on the islands. Right behind are New Zealand, with about 2,000 species and 81 percent endemism, and New Caledonia, with about 3,250 species, of which 76 percent are local. These three islands are in a class by themselves globally, along with aforementioned New Guinea. Then there are the Juan Fernández Islands, at about 200 native plant species and 62 percent endemism; Cuba, at 5,900 species and 46 percent; Hispaniola, at 5,000 species and 36 percent; and, bringing up the rear, the Galápagos, at 700 and 25 percent; Jamaica, at 3,250 and 23 percent; and Puerto Rico, at 2,800 and 12 percent. The explanation for this pattern of rarity among islands, using endemic plants as a rough indicator, would likely reflect distance from a mainland, topographic relief, variety of habitats, and in the case of continental islands, the age of the island since separation. On these islands, some of the endemic plants might be superabundant locally, not rare in number at all, though still rare by the criterion of range. Others are rare not only in range but also in number.

Other members of the expedition were concentrating on vertebrates other than birds. One evening the Kwerba guides reported a brief sighting of a golden-mantled tree kangaroo, one of the rarest mammals in New Guinea and one of the most uncommon in this group of marsupials.

The golden-mantled is a handsome six kilograms of silky chestnut-brown fur offset by light underparts and a yellow wash on the neck, cheeks, and feet. A double golden racing stripe runs down the back. The yellow and maroon furry tail is almost one meter long. The naturalists' find was the first sighting of this animal in the Indonesian portion of New Guinea and represented only the second known site of occurrence. Tim Flannery, a world authority on New Guinea mammals, first described this species in 1993; it had been sighted across the national border to the east, in Papua New Guinea.

We don't usually think of kangaroos hanging out in trees, but ten of New Guinea's native kangaroo species prefer an arboreal life.

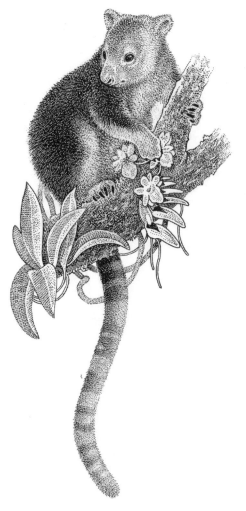

*Golden-mantled tree kangaroo (*Dendrolagus pulcherrimus*)*

Tree kangaroos will never match the acrobatics of monkeys or gibbons, but tree roos are nevertheless skilled leapers and skydivers, able to fall twenty meters to the ground without breaking a limb. Once on terra firma, tree roos lack the athletic grace of a wallaby or wallaroo, but they can still scamper at speed. The arboreal lifestyle, however, is a comparatively recent evolutionary move. Not unlike whales, whose ancestors were land mammals that headed back to sea, tree kangaroos evolved from ground-dwelling kangaroos that in turn descended from arboreal opossums.

Why the kangaroos reclaimed the treetops remains a mystery, but one reason might be that in the rain forest many of the most delectable leaves are above ground level. Tree kangaroos love to nibble epiphytic orchids, among other canopy-dwelling plants. New Guinea's tree roos feed both in the canopy and on the ground in the early morning, thus taking advantage of the absence of primates on this side of Wallace's Line to garner the fruiting displays of trees while also occupying the niche that a ground-feeding herbivore might hold elsewhere.

Flannery also noted when he was in PNG the devastating effect of subsistence hunting, especially for species that are naturally rare and have a very circumscribed range. He suggested that one can predict the extinction dates of rare birds and mammals on the basis of the arrival time of Christian missionaries. Before missionaries came, mountain clans respected strict taboos on hunting. Some places were off-limits, such as where spirits dwelled; some species benefited from taboos on hunting during the breeding season as well. These bans honored traditions but also ensured that hunted species with low reproductive rates could recover. When priests arrived, they urged converts to ignore taboos and other pagan beliefs; overhunting was one result.

The team kept their eyes out for more wildlife sightings. The absence of local hunters in the upper Fojas meant that the region might be ideal habitat for another species seldom observed: the long-beaked echidna, New Guinea's oddest and least studied mammal. On one of the last few nights of trail surveys, frog expert Steve Richards spotted an echidna that was slowly trundling down one of the survey paths. One of the local men put the rather helpless creature in a large bag and toted it back to camp for the others to see.

The long-beaked echidna is the largest of the four echidnas. Although more commonly known as spiny anteaters, echidnas have only an ancient relationship to the anteaters of South America. Echidnas and their distant relative the duck-billed platypus defy the belief that mammals are unable to lay eggs. Egg laying reveals these primitive mammals' primordial relationship to lizards and birds and earns these species their own order of mammals, called the Monotremata. Had the island's hunters frequented these uplands before the Beehler expedition, however, echidnas would have vanished, their meat being highly prized.

The highlanders may be among the sharpest-eyed hunters in the world. Their skill evolved from necessity, as dietary protein is hard to find in this region. Their knowledge of local natural history is unparalleled and their concept of taxonomy decisive, if nothing else. The late Ian Craven handed local hunters an illustrated guide to the birds and mammals of the Arfak Mountains, west of the Fojas. They laughed and pointed excitedly at many of the species found in their area but stared blankly at other common residents. Ian was puzzled at first: he knew they had seen, and had a local name for, everything in the guide. As it turned out, the hunters' binary taxonomy of approval or silence had little to do with presence versus absence or common versus rare; rather, their classification hinged on the Kingdom of the Tasty and the Kingdom of the Inedible.

Life in the fog and the burgeoning quagmire: Bruce and his team could add another dimension to the isolation they felt. Sometimes the mist was so thick that the upper Fojas became their own world in the clouds. And then there were the incessant downpours. Bruce's field notes took the voice of an adolescent sent to summer camp for the first time, trying to convey his joy in the outdoors to his parents while simultaneously scaring the daylights out of them. "Camp grows ever more horrific with continual rain and mud. Wear high rubber boots all day every day. Forest is superb! Have added 12 new bird species to Foja list!"

Bruce's expedition was confirming the impression Jared Diamond had formed in 1979 on his first visit—this was indeed a lost world. Every day brought a new natural history revelation, as if Bruce had pinpointed with his GPS the lat-long of paradise for a naturalist. The Fojas also offered a new perspective on seclusion. For exiles such as Alexander Selkirk, Robinson Crusoe, and even Napoléon Bonaparte, a remote location was viewed as punishment. But in the mountains of New Guinea, the tropical Andes, and the eastern Himalayas, geologic events have led to the separation of populations and adaptive radiations. These phenomena have bestowed upon the natural world a menagerie of rarities—birds and mammals and plants and butterflies and beetles so extraordinary, gifts over evolutionary time of natural isolation.

After two weeks, though, the expedition was drawing to a close. Food was running out, and the hard-won research permit was about to expire. Bruce's diary: "4 Dec. The bog turns into a lake again! A problem if helicopter comes when bog is a lake! Tons of rain falling and trails are appalling." The weather was indeed getting worse, but the discoveries kept coming, right up until the last day of the expedition.

There remained a small problem—the ability of the helicopter to fetch them safely. Bruce's notes summed it up with his usual gusto: "7 Dec. Heli arrives at 4 PM (was supposed to arrive at dawn!). We take 45 minutes to get out with enveloping cloud stopping our escape. Three tries before we actually get out of the cloud and to the clear where we can see our way down the mountain—very frightening . . . Arrive at Kwerba at dusk. The others must spend night on the mountain while we luxuriate in warmth and washing in the stream with soap!"

~

Bruce's expedition could take pride in its place in the annals of scientific exploration. But the Foja expedition had missed the golden age of natural history. That period reached its apogee in England about the middle of the nineteenth century. Had Bruce's group steamed up the Thames in 1837, they would have known triumph: handshakes all around from Darwin, Wallace, Huxley, and the officers of the Royal Society, perhaps even knighthood. Instead, the expedition's leader was met by his family and a faithful golden retriever. Bruce immediately returned to his desk at Conservation International. Proposals needed writing, spreadsheets checked, and the bureaucratic machinery jump-started after his hiatus.

Weeks after the expedition's return, Conservation International's public relations department issued a press release about the team's successes under this tantalizing headline: "Scientists Uncover Biodiversity Trove in a 'Lost World' in Western New Guinea."

"I was as close to the Garden of Eden as you're going to find on Earth," Bruce remarked. And with that sound bite, for about a fortnight, Bruce Beehler may have been the world's most famous field biologist or, thanks to his dramatic phrasing, the most quoted. His reference to Eden echoed through the popular press. Requests for interviews poured in. Beehler obliged and offered more statements ripe with imagery: "There was not a single trail, no sign of civilization, no sign of even local communities ever having been there."

Most reporters didn't care about Beehler's spectacular faunal finds; a new honeyeater meant little to them. It was his words that painted a serene vision, one that offered a different, uplifting message, as high and rare as the Fojas. In contrast to the daily stream of mayhem in Iraq and Afghanistan, they could report that an Eden still exists on Earth. Bruce avoided mentioning, though, that what passes for paradise to a biologist is typically a miserably wet place for most mortals. Better to stick with the sound bites.

What, in retrospect, can the Beehler expedition tell us about rarity and abundance? One of the first insights involves the distinction between geographic rarity and population rarity. Most of the endemic species of the Foja Mountains exhibit geographic rarity—tiny global distributions. In their natural habitats some of these species may be, like the golden-fronted bowerbird, easy to locate and observe but may nevertheless trick the biologist into treating them as "common" when they are really just "obvious," an important distinction. The real revelation for some species, such as the six-wired bird of paradise, is how narrow their elevation distribution is—just a vertical sliver of the mountainside—thus making them globally rare because of their restricted range. By contrast, the golden-mantled tree kangaroo is twice rare—it is both geographically rare (confined to a few mountain uplands in New Guinea's northern coastal ranges) and population rare: hard to find even in its favored habitat.

Birds of paradise of one species or another once inhabited all of the island's area below timberline. But their distinct distributions reveal a clear pattern of rarity. Ten species are rare because they are endemic, confined to tiny mountain uplands or offshore islands. Some, such as the black sicklebill and pale-billed sicklebill, are naturally rare or uncommon, with very low population densities. They are the largest members of the bird of paradise family (the Paradisaeidae), and these outliers tend to be widespread but nowhere common. Others, including the yellow-breasted bird of paradise, appear to prefer very specific habitats, being absent from some sites and found in others.

Although the main purpose of the Foja expedition was not to parse rarity but to discover new species and recover lost ones, it nevertheless confirmed that even in a place without a historical presence of humans, rarity is a common phenomenon. In this region, with its tiny human population confined to the lowlands, birds of paradise are not threatened by hunting. Thus, we can see their populations at what biologists call carrying capacity—densities one would likely see that reflect what the habitat can support. From Bruce's field notes it became clear that some, including the manucodes and the lesser bird of paradise, are naturally common within their range and easily encountered. Others, such as the pale-billed sicklebill, are uncommon and elusive. By contrast, the remarkable black sicklebill, which has been hunted out in many parts of New Guinea, turned out to be present in fair numbers in the

Foja uplands, where no hunting takes place. The big surprise was the absence of the superb bird of paradise, which is perhaps New Guinea's most common and widespread mountain-dwelling species. It is known in virtually every mountain range in New Guinea except for the Fojas.

The last observation is truly puzzling and requires further study. But the other sightings, or lack of them, show that even in perhaps the most remote spot in the tropics, the rules of rarity still apply. Many species—be they birds of paradise, tree kangaroos, frogs, butterflies, or palms—naturally occur at low population densities and hence are rare. Viewed the other way around, there may be something about the very nature of rain forests that prevents some species from becoming superabundant. The Fojas are rich in species, so perhaps the sheer number of species in the ecosystem leads to a novel mechanism at work—a kind of diffuse competition in which the interactions of many species keep others in a limited range or at a low density. We will revisit this potential structuring force in the Peruvian Amazon in the next chapter. For now, it is simply enough to note that ecological interactions in the New Guinean rain forests seem to keep many species scarce.

In further conversations with Bruce, I turned to a behavioral trait that intrigued me: the lack of shyness shown by the wildlife. Could it be a result of evolving in a region with no large terrestrial predators to stalk them or humans to hunt them? Bruce confirmed my guess of the relative dearth of predators in this upland forest. So, if large and midsized spotted cats never cross Wallace's Line and aerial predators are uncommon, then looking up or around is a distraction from dancing. Bruce concluded, "It makes the forests of New Guinea something of a peaceable kingdom, friendly to the birds of paradise and bowerbirds." It may seem as if Bruce slipped back into sound bite mode again, but not entirely. Nature is replete with starvation, disease, intense parasitism, and death from typhoons or drought, even in the Fojas. His point was simply that here was a place where fierce predators of these birds had less of an ecological and perhaps evolutionary influence.

In the Foja Mountains, the birds seemed to be common within their narrow range and tame, whereas the mammals appeared neither common nor tame. Perhaps the explanation is that some traits selected for in response to encounters with human hunters had persisted even when such wildlife populations dispersed into new areas. The range of these species of mammals included areas hunters frequented. Some, such as the six-wired bird of paradise, changed, reverting to natural boldness in isolation from humans; others, such as tree kangaroos, which remain shy, seem to have been more resistant to evolving to take advantage of the absence of human predators. Bruce had invoked what scientists call "the ghost of evolution," that is, in this instance, the "echo" of past influence of New Guinea–wide hunting pressure continuing to resound through time, even in the absence of the original hunting pressure, at least for some of the mammals. "What we are actually observing is natural selection on the local scale in real time, not over thousands of years. Only the stealthy or wary species survive the hunter's bow, snare, or shotgun when wildlife meets humans for the first time. It's why some birds of paradise disappear right away near villages."

Further analysis of some of the other taxa supported the findings on birds of paradise, bowerbirds, and tree kangaroos. New Guinea is one of the global centers of diversity of palms, with about 145 species plus 5 new ones discovered by the Beehler team. What were the patterns of rarity and abundance among palms or orchids or other vascular plants? Was the pattern here similar to what we expect in other tropical forests: a few abundant tree and understory shrub species and many other species represented by a few individuals? Was this true for butterflies as well? Brother Henk van Mastrigt confirmed that for the groups of butterflies and moths he collected, the pattern held. Many frogs, too, showed the same basic pattern as the birds and mammals of the Fojas: a few common species with very limited distributions, with a number of other frogs that were range limited and had low population density, in comparison with a few Fojas species that were widely distributed across the northern mountain ranges of New Guinea.

The combined data on mammals, birds, and amphibians led to some important conclusions. First, the Foja range clearly emerged as one of the world's centers of mountain-dwelling species found nowhere else on Earth. Second, its fauna is a wonderful example of localized speciation fashioned by geologic processes. Isolation and speciation, again, are engines of rarity. Of course, the Foja expedition still leaves lots of blank pages to fill in about nature without humans. The costs and logistics of an expedition limit researchers such as Bruce Beehler to short-term visits, even though long-term studies are required to unravel the patterns of rarity. Until permanent research camps are established, insights into patterns of rarity in places without humans will be mined slowly. Thus the paradox of accessibility: if upland forests were more open to biologists, we might know more about rarity and abundance for all the vertebrates, yet if they were more reachable, the vertebrates would likely be gone. Two return visits to the Fojas in 2006 and 2009 by Bruce and colleagues also offered new data. To cite one example, the scientists encountered the golden-mantled tree kangaroo only three times in three trips—that is rare by any measure. On the second trip they never located that species, nor did the Papasena hunternaturalists who went out in search of it. The echidna was not seen on the final two trips—not once.

Places such as the Foja Mountains are already quite rare globally and will remain so. The range covers more than 3,000 square kilometers of untouched old-growth tropical forest sans humans. While such are the places of naturalists' dreams, the Foja natural experiment is impossible to replicate in many places in the twenty-first century. One of the best examples of what might be called "restored isolation," though, comes from across the island. From

1998 to 2008, Jared Diamond spent time surveying an area in Papua New Guinea's Kikori River valley, where Chevron has built a 171-kilometer pipeline to ship oil from an inland mountain valley to the coast. The pipeline road was gated to keep intruders out of this vast wild landscape. The result was a reverse experiment in recovery of the native fauna. Diamond noted species that had become common again, such as the southern cassowary, hornbills, birds of paradise, some of the giant pigeons, parrots, tree kangaroos, and hundreds of species of the forest interior, all of which normally disappear once hunters gain access. As much as the guns-and-fences approach to conservation is dismissed by so many these days, the Chevron Kikori experiment illustrates how once rare species can recover with strict protection. Restricting whole landscapes, as in Kikori, however, is neither practical nor ethical and certainly is not affordable unless financed and policed by, for example, a major multinational energy company.

In truth, the Foja Mountains are well guarded by the Kwerba and Papasena, who control access. This range is their patrimony and their most important source of wealth—timber, pure water, meat, medicinal plants, even bird of paradise feathers—and they will not be giving it away to outsiders any time soon. After Bruce's helicopter lifted off the lake bed on his second visit to this remote domain, the montane forest would return to its splendid state of isolation. The visitors had gone home. The bowerbirds would continue to dance around their bowers. Nearby, the male six-wired would shake himself into a joyous blur, and the uninhibited displays of other birds in this paradise would resume without a human audience.

Chapter 3

A Jaguar on the Beach

*T*ROPICAL RAIN FORESTS OFFER stunning exhibitions of animals and plants. Many artists have tried to portray this richness, none better than the French painter Henri Rousseau. His canvases, filled with curious primates and fearsome wild cats, grace the walls of leading museums, and reproductions appear on the covers of ecology textbooks. Despite Rousseau's lack of formal scientific training or travel to equatorial regions, his paintings capture many of the central themes in modern tropical ecology. A biologist wandering through a Rousseau retrospective would marvel at how the Parisian anticipated such key topics as the importance of predation by large cats on large plant-eating mammals, the abundance of plants whose seeds are animal dispersed, and even a visual hint of the rarity of tropical forest dwellers.

To understand rarity in nature, whether as an artist or a biologist, one of the best places to look is in the tropical belt. The Amazon and Congo basins, Southeast Asia including Sumatra (Indonesia), and New Guinea are the four largest expanses of rain forest; along with some smaller regions, they hold more than 60 percent of the world's known species—crammed into less than 5 percent of Earth's surface. Because these rain forests are incredibly rich in species, they contain unusual numbers of rarities. The Foja Mountains of Papua Province, Indonesia, on the island of New Guinea, serves as a great starter location for understanding how rarity is created through extreme isolation on mountain chains. The Amazon rain forest, the next locale on our journey, is a vast, low-lying region where native tree species and a wide variety of vertebrates that inhabit them—carnivores, monkeys, and macaws, to name a few described here—illustrate another crucial type of rarity to consider: species that occur at a very low density but over a wide range. Three of these great reservoirs are mainland rain forests. They offer an important missing element in the mix of rarities that is absent in the New Guinea fauna—top predators, especially the large cats. These cats stalk their prey on the rain forest floor: jaguars and pumas in the Amazon, leopards in the Congo, tigers and clouded leopards in parts of Indonesia and Indochina.

One of the most cat-rich rain forests on Earth is a nearly unbroken belt of wilderness in the western arc of the Amazon basin. In the secluded Madre de Dios region of Amazonian Peru, we can examine how and why individuals of the wild cat family and other terrestrial organisms inhabit great expanses but are relatively distant from one another. In this remote Amazonian locale, enclaves also exist with no evidence that large cats have ever been hunted. In such a sanctuary, we can see how rare rain forest mammals and birds, such as jaguars and pumas, even at very low densities, influence the species-rich domain over which they rule.

On a sunny morning in August 2009, in the very heart of the Madre de Dios region, I joined the husband-and-wife team of George Powell and Sue Palminteri to survey a stunning rain forest panorama from the top of the canopy-viewing tower in Peru's

Manu Wildlife Center. Some fifty meters above the ground, I wondered what Rousseau would have chosen to portray in the scene before me. Surrounding the tower was perhaps the world's richest natural arboretum, a forest holding one-tenth of all bird species and enough as yet unnamed invertebrates to fill several natural history museums. Acrobatic spider monkeys and their slow-moving relatives, red howlers, appeared below. In a neighboring fig tree, brightly plumed trogons and barbets decorated the branches. Flocks of tricolored scarlet macaws flashed by at eye level, penetrating the forest with their cacophonous squawks. For these canopy residents—monkeys, macaws, and thousands of others—this is the sweet life on the breezy, green roof of the rain forest. The chance to join them in this restorative stratum, even for an hour, is a relief from the claustrophobia of the dark, dank forest floor. For the elevated naturalist, the exhilarating climb to the top serves as a refresher course—a moment to ponder when our ancestors, too, were arboreal.

The tropical rain forest is full of common species such as ants, nematodes, and fungi that contribute heavily to the unseen machinery of an intact rain forest, yet it is a bountiful Kingdom of Rarities as well: Many living things in this landscape, from the tops of emerging giant trees to the thin soils covering their roots, exhibit several forms of scarcity. Some insect species exist only in the canopy of a single tree species, but there they mass in numbers. The giant trees orchestrate life in the Amazon and create a three-dimensional stage for millions of smaller organisms and staggering diversity, too: a single tree may hold more ant species than are found in the entire British Isles. Yet the distribution of these trees is quite different from that in England and western Europe, where a few elms, maples, beeches, or oaks may dominate a large patch of forest. In the Amazon, particular trees, such as mahogany and members of the Brazil nut family, may range eastward across Brazil but be spread out as individuals, occurring maybe only once every few hectares, sometimes even more dispersed. The jaguars and pumas that wander below their crowns and several of the monkeys that live in them also range widely but are thinly distributed. The macaws that fly across the canopy also appear across much of the lowlands but are locally scarce.

The unusual flash of brightly colored birds in an all-green realm triggered a raft of intriguing questions that had been lingering in my mind for years: Why might such species be rare? And why are rare creatures as different as macaws, monkeys, cats, and rain forest trees so similar in the character of their spacing patterns? Was one single factor the cause, transcending whether an organism was covered in fur, feathers, or leaves? And why do biologists believe that some of the scarcest species exert a tremendous influence on the workings of the rain forest?

Up in the Manu viewing tower, the distant sound of communal throat clearing caught Sue's attention: the hubbub was a troop of capuchin monkeys expressing their displeasure at an intruder lurking nearby. Perhaps a jaguar had walked past and settled along the riverbank or a forest eagle or falcon had passed overhead. "As long as the capuchins stay hidden in the branches, they're generally safe from aerial predators," Sue related. "But they're scared to death of an opportunistic spotted margay cat that could climb up after them." At all costs, capuchins and other monkeys avoid the ground, where they would be an easy mark for a jaguar. This largest of South American cats is known by its local name, *yaguara*, which means "to kill with one pounce."

Consistently locating jaguars was central to George's research project. He sought to answer one of the most pressing questions of tropical biology: How much is enough? He wanted to know how much habitat area was needed to conserve jaguars, pumas, peccaries, tapirs, and other locally rare Amazon species. Up to now, scientists could only guess or shrug in response.

Sue had spent the previous five years searching these forests for wild primates, the little-known saki in particular, so she knew all about the wariness of monkeys. The bald-faced saki, part of an unusual-looking group of primates, the pithecians, is distinguished by its long, coarse fur, a pelage seemingly out of place in the tropical rain forest but quite useful as camouflage, an important feature for canopy residents that weigh only two kilograms. Their very long, black-and-white mottled hair not only makes these monkeys appear twice their size but also resembles the shape, color, and texture of the branches of their home in the lower canopy. Their unusually bushy tail only adds to the strange silhouette, although it is not prehensile like the tails of larger-bodied New World monkeys. Rather than hang from tree branches as do some other primates, they scamper over canopy limbs like giant squirrels. When they reach a tree laden with unripe fruit, sakis use their massive (for their size) canines to pry, rip, or split open the often hard-coated fruits to reach the soft young seeds. But few primatologists had ever seen bald-faced sakis for very long, let alone studied them. Even their weird vocalizations seemed like part of the "camouflage" adaptation: their birdlike sounds, on top of their secretive behavior, contributed to Sue's own ecological name for this group: stealth monkeys.

Sue and her associates had been the first to acclimatize troops of bald-faced saki monkeys to human presence and watch them every day for hours on end. Eventually, the primates became comfortable enough to act without visible concern about their presence. That allowed her to focus on the contributions of her sakis to the rain forest ecology. Sakis seemed to be everywhere but nowhere at the same time, she noticed. A distribution map of the saki indicates the species is found throughout Madre de Dios, but when one tries to find them, the monkeys' presence seems irregular and unpredictable, what biologists call "patchy." Why should that be so when all of that forest looks so similar to us?

Monkeys may be hard to spot, but they glow like neon compared with jaguars. Few veteran tropical biologists have ever met a wild jaguar; according to those who have, a prolonged encounter is usually a once-in-a-lifetime event. The cat's exquisite pelage of black rosettes on an amber background stands out in animals on display in zoos and on wall calendars, but that coat camouflages them in the dappled light of the forest understory.

Many biologists who attempt to study rain forest mammals eventually give up and shift to some other project. It is not for lack of trying. Rain forest mammals shrink from plain view. Many are nocturnal and diminutive; they scurry under leaves or, like the most abundant group—bats—roost in caves or shelter under dense boughs until nightfall. But even large vertebrates that are active by day are expert at staying out of sight, especially where hunters stalk them.

George's two target cats—jaguars and pumas—are among the most wide-ranging mammals in the region. The jaguar roams from Arizona to Argentina, but unlike many tropical rain forest denizens it can live in desert conditions. Like most large predators, however, it always occurs in low densities. The puma, also known as the mountain lion or cougar, can be found from the Yukon south almost to the tip of South America in the mountains of Tierra del Fuego. Like the jaguar, the puma has many ecological addresses, from the edge of arctic tundra to tropical wet mangroves, but never exists in large numbers. Unlike jaguars, pumas are one of the most studied of all large cats, except in the Amazon. Here, their ecology remained a mystery until George became intrigued by them.

How do the two big cats fit together in this ecological puzzle? "I began to wonder," George told me, "do the smaller pumas share space with jaguars under the rain forest canopy? Do they hunt the same areas for prey?" Reduction of prey by another top predator would only boost the total area needed to support a pair of breeding jaguars, say, and make the answer to "How much is enough?" frighteningly large. Of course, from the perspective of a territorial male jaguar, even one nearby male jaguar or puma might be too close for comfort.

Since these questions all related to what rarity is, I began to wonder about the relativity of our definition and how we understand the concept. Do jaguars and pumas even qualify as rare species? Some biologists point out that when it comes to rarity, both of these big cats, as well as the lion and the tiger, have such a widespread distribution that it may compensate for their low density wherever the species occurs. Rarity, they argue, really depends on the scale you are talking about. Other scientists argue that explaining the rarity of jaguars and pumas is as simple as looking at a food pyramid, as Paul Colinvaux summarized in his aptly titled 1979 book *Why Big Fierce Animals Are Rare*. It takes a lot of prey animals to support a healthy population of large carnivores. Biologists refer to different levels of food gathering by organisms as "trophic levels," which can be visualized as a pyramid with the largest meat eaters at the apex, followed by fruit and insect eaters, then plant eaters, and finally the plants and fungi themselves at the base, supporting the whole structure. Species that occupy the top of the food chain are always scarce, whether on land, in lakes or streams, or in the oceans. Thus, in the rain forests of the Amazon, the Congo basin, or Asia, the largest flesh eaters, as dictated by the laws of thermodynamics and energetics, have to be much fewer in number than the large herbivores they eat. But how do the causes and consequences of rarity vary as one moves down the tropical food chain, from big, fierce cats, large herbivores such as deer and tapirs, monkeys, and macaws to the trees themselves, which give the forest its three-dimensional structure? The work of George, Sue, and other colleagues has brought these issues to light.

Knowing something of the density of the cats' prey could help George to better understand patterns of rarity among those that stalk them. Meet the forty-kilogram white-lipped peccary, a shaggy, piglike forest specialist and a primary prey item of both jaguars and pumas. Unlike the jaguar, a peccary moves in the company of hundreds of its family members and associates. Peccaries are constantly on the go and travel long distances over the year; it is still not clear whether they are nomadic or migratory. Unlike cats, peccaries make their presence obvious. Rooting peccary groups leave conspicuous pockmarks and furrows in the soil and announce their presence with grunts, screams, and tooth clacking. The sound track is accompanied by an overpowering barnyard cologne, an especially pungent odor that peccaries emit when excited.

George and Sue chose the remote Madre de Dios area of the Amazon, and the Los Amigos Biological Station in particular, for their studies. Los Amigos offered a rare chance to study wildlife in a hunter-free zone. Somewhat like the Foja Mountains of New Guinea, Los Amigos can be used as a control site for other studies conducted in places where hunting is intense. At Los Amigos, George had the opportunity to study predator and prey populations, and Sue could study monkey troops, where their subjects' numbers and behavior were less influenced by human contact. For example, high densities of peccaries where they are neither shot nor snared might offer different insights into the wanderings and densities of cats from findings in locations where prey has been decimated by hunting, as it has been in a forest area of Madre de Dios near Puerto Maldonado, southeastern Peru's regional population center. These contrasting sites permitted George to study how wildlife responds to intrusion and even invasion. Near town, machetes and chain saws were carving up this once isolated stretch of the Amazon. Tens of thousands of gold miners were digging in the forest and dredging the riverbanks. Construction was under way on the new Transoceanic Highway, linking Brazil with the Pacific Ocean. Completed in 2010, it now connects southeastern Peru to the outside world—and it fragments what was one of the largest expanses of undisturbed rain forest. Thousands of poor villagers from the high Andes had already settled along the way in search of work, and more were arriving daily to seek their fortunes in the Amazon lowlands, as had the forty-niners in California 150 years before. The human migration is ongoing, despite the history of nearby Rondônia, Brazil, where in the past few decades thousands of settlers have staked their dreams on a piece of Amazon geography only to find livelihood unsustainable on its poor soils.

Map of the southeastern Peruvian Amazon

The shrinking of the rain forests because of human settlements introduced a new set of research questions for the husband-and-wife team. How had loss and fragmentation of rain forest and new levels of human influence, especially hunting, affected answers to the initial questions of how much is enough, for jaguars, for pumas, and for sakis? Given the pace of forest degradation, finding answers to these questions is urgent.

For the past four years, George had contracted with a locally run air taxi company to help him track the animals he and his field team captured and fitted with radiotelemetry devices. It was time for the near-monthly flight to pinpoint the animals' locations, and George had offered to take me along. At the grassy field that served as an airstrip, George greeted his pilot, Jonathan Schmidt, and quickly attached his tracking gear to the plane.

George was always keen to apply a new technology or to tinker with older versions. When his research began four years earlier, in order to learn how many jaguars lived in this area and how much of which habitats they used, George and his colleagues placed automatic cameras known as camera traps in the dense forest to document the presence of species, such as jaguars, that shun detection by human eyes. A camera is set up along an animal trail, and whenever a moving subject passes in front of it, a sensor is triggered and— *Click!*—a photographic record is made, with an electronic date and time stamp. The results can be breathtaking. With the use of hidden cameras, species that were once ghosts—never seen by anyone or absent for decades—suddenly reappear.

Remote camera traps also enable estimation of actual densities of species that are almost never seen. Biologists have adapted statistical methods to derive numbers from photographic records of known individuals. Luckily, jaguars are perfect for this system because their spot patterns offer unique identification for each cat. The trick, though, is to set up opposing cameras to photograph each side of the passing jaguar (or other spotted or striped cat) because the spot pattern differs on the left and right flanks. If George and his team saturated the forest with enough camera traps placed at proper intervals and covered a sufficiently large area in a relatively short time, they could obtain a density estimate.

Even with the cameras, however, George's question could not be fully answered. The method depends on the area studded with cameras being larger than the home range of the study animal. Yet George's initial camera trapping grid, an area about as large as 10,000 soccer fields, proved too small to contain the entire home range of a single jaguar. In fact, one male used an area eight times larger! At best, the camera trapping effort offered an indication of jaguar presence.

To learn more about their daily movements, George had no alternative but to start catching jaguars and fitting them with tracking devices so that he could map the movements of each individual. The jaguars proved elusive. Adapting techniques from other studies, George and his team set large live traps—baited with pigs, dogs, and chickens, each housed safely in a separate compartment of the trap—in different parts of the forest.

Despite many attempts, these baited traps captured very few jaguars. George called in some expert cat catchers, who suggested that recordings be played of the cats' breeding calls and of vocalizations made by their prey. This decoy did the trick. Within a year, George and his team had more pumas and jaguars radio-collared than had ever been accomplished before in the Amazon.

In a region with no roads, however, tracking radio-collared jaguars on foot or by boat turned out to be futile; even when collared animals were nearby, the dense forest swallowed up the signals— the jaguars might as well have been a hundred kilometers away. In an airplane flying above the canopy, though, the signal was loud and clear, and George could locate all of his collared animals each time with a few hours of flying.

"Let's go find some jaguars!" George shouted to me over the engine noise. He gave a thumbs-up to the pilot, who then set the Cessna hurtling down the grassy, water-soaked runway in hopes of clearing the tall forest wall that surrounded us on all sides. It was a joy and a relief to rise above the living curtain of trees into a beautiful sunrise. Below were trees in a thousand shades of green as far as the eye could see, dotted by occasional ones with young, bright red leaves and a few with crowns awash in yellow or pink flowers.

What appeared to be a single type of forest, however, was many forests in one. Meandering rivers are a trademark feature of the Amazon basin and shape the extent of floodplain forest, where tree species must cope with periodic inundation. The high disturbance levels contribute to floodplain forests being less rich in species than the more stable upland forests, beyond the reach of seasonal floods. These upland stands are among the most diverse forests in the world. A thousand shades of green may not be far off as an estimate: perhaps as many as 1,000 woody tree and vine species populate the area, though the tree species occur at extremely low densities.

In a clearing along the banks of the Madre de Dios, the buildings of the Los Amigos Biological Station gleamed in the morning sun. As the plane circled, George found his first radio-collared jaguar. Locating jaguars by plane may seem romantic, but it was really a throwback to the old ways of wildlife tracking. The schedule of determining one location a month for each collared cat would yield few answers to George's overarching questions. Frustrated but determined, he considered satellite telemetry using global positioning system collars. But even with expensive GPS collars, he concluded, reception would be possible only when the collared animals ventured into an opening, such as a rare natural clearing or a river sandbar.

The jaguars we were locating now reflected George's most recent solution to the problem of tracking in such dense vegetation. As the plane swung toward the Tambopata River, the signals of two more jaguars pinged in the headphones. Since 2006, some of the jaguars below us had been carrying two complementary signaling devices, one traditional—which was giving George the location he heard in his headphones—and the other a piece of wizardry. Mounted on the standard VHF radio collar was a TrackTag, a device invented by the Scottish engineer Peter Brown that obtains a fix in about twenty milliseconds—a significant improvement over normal GPS collars, which take at least ten times that long. Using this technology, George collected thousands of fixes, rather than the few he could obtain using VHF collars, over the course of our three-hour flight on that first of our two-day jaguar reconnaissance.

"It's really pretty ingenious," George said later. "After the Track-Tag picks up the signals of the GPS satellites, it stores hundreds, perhaps thousands, of fixes on a chip rather than attempting to transmit the location back to us via a communication satellite, as other GPS tags do. After we recapture the animal, we swap the TrackTag and download its data. But if we don't recapture it, we can program the collar, with VHF and TrackTag, to drop off the animal. Then I can fly, locate the dropped collar using the VHF device, and send my field crew to that point to retrieve the Track-Tag." After returning it to Peter Brown, George would receive a computer file with all locations that the jaguar or puma had visited while wearing the TrackTag.

"Couldn't you learn the same thing from the VHF signals you pick up by flying each month, even with fewer locations?" I asked. "Take a look at this," George responded, indicating an image on his laptop. Since the beginning of the radio tracking, he had been religiously plugging his jaguar locations into a database. On the screen were the location points, collected during all the flights, for Paya, a female that had recently been killed by another jaguar. The seventeen VHF fixes George had for Paya, when connected, outlined a large triangular home range of about 230 square kilometers overlaid on a map that covered vast sections of upland forest in addition to the Tambopata River.

"Now watch," George said. He keyed in a few commands and displayed the TrackTag locations—literally hundreds of yellow dots—on top of the triangle created by the seventeen VHF fixes. "Quite a revelation, huh?" he said with a laugh. The TrackTag locations, recorded during the same period from the same animal, showed exactly the opposite pattern: Paya hugged the fifty kilometers of the river system lying within that large triangle. "She actually avoided the upland forests!" George exclaimed. "I had a hunch that the sparse VHF locations might be misleading us about how jaguars use Amazonian forests. Now we know."

When George uploaded TrackTag results from other animals, it was clear that Paya was not the only big cat to follow that pattern. Other females largely avoided the much more abundant uplands and patrolled the riparian forest instead. Although the TrackTag evidence was indisputable, the behavior was mystifying, since rivers have been the center of human development in this region for hundreds or even thousands of years. Why did the jaguars live so close to the water's edge? What was attractive enough to override the real threats posed by a growing human presence? In the pioneer days, the threat had been boatloads of pelt hunters. Today, the threat is thousands of gold miners and their followers who provide them with food, lodging, and pleasures. And still the jaguars hang on.

As I write, those questions remain unanswered and George still labors to crack the mystery. The evidence so far, however, casts doubt on his original assumption, that the jaguars have a taste for river-dwelling food items. Using modern DNA-sequencing techniques to analyze the jaguars' scat and identify their prey, his team has found no sign of fish or turtles or caimans. Instead, white-lipped peccaries are the primary prey. Tracking results for those animals showed that the peccaries range widely throughout the floodplain, so a steady diet of these wanderers gives no obvious clue to why the females are so attracted to river edges. Perhaps the question will remain unanswered until a researcher figures out how to add digital cameras to the TrackTag equipment in order to monitor how and where the female jaguars capture their prey. One theory holds that it is easier for cats to kill peccaries along the flat riverbanks than in the broken terrain of the uplands.

In the meantime, George has looked at the tracking data for the male jaguars to see whether they follow the same patterns. It turns out that males tend to wander more widely over the floodplain than do the females, perhaps moving from female to female in search of breeding opportunities as large cats are typically known to do. George speculated that the big males, almost twice the size of females, can more easily bring down the tusk-laden peccaries—which themselves are nearly the size of female jaguars—with less dependence on the element of surprise. This may support the view that females prowl the riverbanks because the dense understory gives them more cover from which to launch an attack. If the males, on the other hand, depend less on short-range surprise attacks, they can focus on their second most important drive, cherchez la femme.

For assembling a picture of jaguar movements, a missing piece was knowledge of the density of jaguars far from river channels, where peccaries are few. To address this, George and his team needed a research site in the upland forest. He found a well-managed timbering operation where hunting was prohibited and obtained the local landowner's permission to use the concession as home base, allowing the team access to the vast uplands. Although this study is still under way, initial data indicate the big cats use larger home ranges in the upland areas, perhaps because of the near absence of their preferred prey, the white-lipped peccary. "My data clearly show that white-lipped peccaries are super common in the floodplain habitats, but rare in the upland forest away from the flooding," George wrote in his journal. "I will be curious to see if plant sampling data show that the extensive upland forest is largely devoid of the large palms whose nuts the white-lippeds depend on. At least that is what we expect."

George's inquiry into how jaguars and pumas share the range, or avoid each other entirely, was only partially answered. His preliminary data indicate that pumas seem to be similar to jaguars in their preference for riverine habitats but with the caveat that there is little or no overlap between male territories. Unfortunately, female pumas turned out to be too small to carry a TrackTag collar. (Typically, researchers try to keep telemetry devices within 5 percent of an animal's body mass, to avoid interfering with its behavior.) Thus, greater insight into how these two cats fit together in ecological space awaits yet another advance in the technology of animal tracking.

One can infer only so much about the causes of rarity by mapping jaguar or puma distributions. Perhaps greater understanding comes from looking across the ocean at another pair of large feline predators: tigers and leopards in South Asia. A common saying among old jungle-wallahs there is that "where tigers are common, leopards are scarce." Tigers are known to kill leopards, and in areas of high tiger density, leopards stay close to the margins of wildlands and closer to human settlements, where tigers tread less frequently. Perhaps pumas, especially the smaller females, are strongly influenced by the movements and ranging of jaguars and stay in areas avoided by jaguars.

Another important aspect of trophic rarity, or the "big, fierce animals are necessarily rare" paradigm introduced by Paul Colinvaux, is prey density. In fact, again using South Asia as an example, researchers have shown that tiger density in a given habitat can be predicted by the prey density estimated for that habitat. Tigers reach highest densities in the riverine grasslands of Asia, over 20 adults per 100 square kilometers, and an order of magnitude lower densities in mature forests, all because large prey animals—deer, wild boars, and Indian bison—do likewise. Perhaps more studies of co-occurring jaguars and pumas and their prey elsewhere in the Amazon will indicate a similar relationship between prey density and predator numbers.

Bolstered by George's TrackTag data, an answer to the initial question, "How much is enough to conserve rare nature as represented by the jaguar?" suddenly seemed more attainable, at least for this region of the Amazon. Individual jaguars in this part of Amazonian Peru have very large home ranges, as much as 400 square kilometers, the data suggested. "If we calculate the jaguar's use of habitat, while accounting for their apparent preference for floodplains and riversides," George commented, "then maintaining genetically viable, healthy populations of jaguars of, say, 500 breeding females will probably require well above 20,000 square kilometers—perhaps between 30,000 and 40,000 square kilometers—assuming a good peccary population." This also assumes that the areas being conserved incorporate extensive riverbank and floodplain habitat. How do the protected areas of the Amazon measure up to this task? Of the more than one hundred protected areas in the Amazon, not one is large enough, and only about a dozen are over 10,000 square kilometers; the average reserve size is only 3,500 square kilometers. Aggregates of protected areas with open passages between them—conservation landscapes, a topic explored in more depth in a moment—will clearly be required.

Studying big cats that were, for the most part, out of harm's way from humans, but not from each other, offered further insights. Each jaguar, male and female, examined by the team provided evidence of these cats' ferocity toward one another. All showed signs of struggle: cuts, old wounds, bitten ears. Several radio-collared jaguars had already perished in the lethal jaws of another: hardly a children's bedtime story, but the reality show of life in the rain forest. A visiting biologist who was familiar with California pumas had a novel view of the violence, suggesting that the frequency of scrapes, wounds, and scars indicated a healthy population as animals fought to maintain access to areas with high levels of prey. A basic law of nature drives this aggression and, ultimately, the wide spacing of jaguars in the rain forest. Perhaps the most important finding was that even where humans have less influence, as in Tambopata, jaguars have a density of only around 4 per 100 square kilometers—rare by any reckoning.

~

While George was tracking jaguars and pumas, Sue and her field team were searching in the same area for the elusive bald-faced saki monkey. Sakis are seed-eating primates, so the "big, fierce animals are rare" theory makes no sense as a descriptor of density. Moreover, sakis live in small family groups of fewer than ten individuals, typically including a single adult male, one to three adult females, and one or more young. At first, Sue figured that the trees whose seeds they eat must dictate the monkeys' locales and limit their range. But careful observation of their behavior and a bit of tree climbing led to a novel hypothesis to explain their rarity.

The answer lay in the architecture of the forest itself. Sue collaborated with Greg Asner, a tropical ecologist with the Carnegie Institution for Science who, together with George, had mapped the aboveground biomass, and estimated from that, carbon density of the Madre de Dios rain forest. The tool they used to collect the biomass data was a laser scanner mounted on a small aircraft. What this innovative technology and revolutionary study showed is that a forest that looks uniform to an untrained eye from the air and even from the ground, looking up at the canopy, is actually quite variable in its three-dimensional structure and biomass. And the sakis saw the forest in a unique way: Sue and Greg found that they could predict where to find these rare canopy-dwelling primates by knowing the forest's biomass—data provided by the laser sensor. The densest branches in the canopy provided aerial concourses for these tree-running primates. In essence, Sue's work showed that the saki monkey was an Amazon old-growth forest equivalent of the famed northern spotted owl of western North America's coastal conifer forests. That nocturnal predator is a habitat specialist, persisting only where the high but open forest canopy still enables it to hunt its preferred prey (flying squirrels) and breed successfully. The main difference in habitat restrictions of the northern spotted owl and the saki monkey is that the patchy distribution of coastal old-growth conifer forests is due to logging. The distribution of sakis in unlogged Amazon forests is a natural pattern of rarity, it seems, because the structure that permits troop movement, the canopy runways of dense, wide branches, is scarce. Sakis may be very common in suitable patches but overall are rare because preferred patches, with the aerial superhighways they require, are rare. The advent of airborne laser scanning in the tropical rain forests of Colombia, Peru, Panama, and Brazil will surely enhance prediction of where rare, dense-canopy primates and other vertebrates might live over vast scales.

∽

When we sat down to breakfast the next morning, we disturbed a clump of fruit flies camped in the bowl of bananas. These short-lived insects feasted on the ripe flesh and skin of fruit. They might spend a fruit fly eternity, measured in days, in a home range no larger than a serving bowl. Across the river, scarlet macaws preened their lustrous feathers in the bright sunshine. Fruit flies and scarlet macaws lead lives that are polar opposites in range, age span, and relative rarity. The noisy birds live for decades and feed on a variety of nuts and seeds. Before George's large-predator project took off, macaws had been one of his main study subjects, along with quetzals and bellbirds, enigmatic bird species of Central America that present urgent conservation challenges. These birds all range widely, and so—in a biological sense—they are surprisingly similar to the big cats. They are "area sensitive," meaning that they need huge expanses of habitat to survive. What can we learn about rarity from their study?

"We've found that macaws can fly hundreds, maybe thousands, of kilometers in search of an available nest cavity," George said, gesturing toward the flock. They seem to know when and where to move to find food, no matter how far away. It is a triumph of avian memory in this vast forested region and possibly a trait under strong selection pressure. Individual birds that could time their food-searching forays accurately enough to find bonanzas of ripening nuts most likely survived to produce more offspring than those that could not.

On the habitat-use continuum from fruit flies to macaws, individuals of most species lean toward the fruit fly end and live out their lives in relatively small areas. Individuals that range over large distances often belong to the very species that networks of protected areas strive but often fail to conserve. So George had begun to focus on another important question: How were the rovings of macaws, jaguars, and peccaries related to specific features of their habitats? He was particularly interested in the distribution of forest tree species that might affect all three species and, of particular interest for the macaws, the spacing of the bonanza nut trees in their range and critical places to nest.

Like the sakis, macaws love to eat the unripe seeds of large forest

*Macaws at a mineral lick, with a jaguar (*Panthera onca*) below*

trees. In fact, they are seen as pests by Brazil nut harvesters, though research has shown they have little impact on yield. With their preferred food supplies dispersed throughout the region and seasonal in abundance, there is no other way for the macaws to live.

Macaws search not just for their bonanza nut trees but also for safe tree cavities in which to raise their young. The blue-and-gold macaws prefer the palm swamps that spring up in old oxbows of the main rivers and nest in the tops of the *Mauritia* palms that grow there. The red-and-green macaws favor as one of their primary nesting trees a canopy genus called *Dipteryx*, which as a fruiting adult, as opposed to a sapling, occurs only about once every five hectares in this region. So, unlike the jaguars and pumas, which prowl at low densities because of energetic laws, or the sakis and other mature forest dwellers that are limited by dense-canopy sanctuaries, the macaws' spaced-out lifestyle at low density is defined by the availability of tree cavities in which to nest, and their preferred nesting-tree genus, *Dipteryx*, is rare.

So we have finally descended the tropical food chain, from the top carnivores to peccaries, monkeys, macaws, seeds, and the species that produce those seeds and give structure to the rain forest. Is the relative scarcity of adult *Dipteryx* the exception or the rule among tropical trees? Are tropical trees any different in the cause of their rarity from the animals that live in or below them? Of the 60,000 to 70,000 species of trees and woody climbers that fill the canopy across the tropical belt, how many are considered rare, and why? To answer these questions, ecologists have to understand the spacing of forest trees and its causes. In Madre de Dios, the problem was that botanists had recorded more than 1,000 tree species in the Los Amigos area alone (to put this in perspective, only 15 tree species are native to the United Kingdom).

The distribution of tropical trees presents the central puzzle of rain forest ecology and, by extension, of one major form of rarity—life at low density over a large range. Explain this ecological brainteaser and many things fall into place. To answer how such rarity is generated requires an understanding of why rain forest trees are so diverse in the first place. From New Guinea to the Amazon to the Congo or Indonesia—wherever you go, you typically find an extraordinary number of tree species in a small plot of land, say one hectare, often with each species represented by a single individual. Why is this pattern so prevalent in the tropics but less so in the temperate zone and boreal forests?

All tropical trees use the same photosynthesis system, utilizing light to convert carbon dioxide and water into sugar and oxygen. Different tree species may "specialize," growing better in different light levels, but the tree species that occupy the canopy (as compared with the understory) have roughly equal access to light and to the available nutrients in the soil. If nature is said to abhor a vacuum, it also seems here to abhor a monoculture. Many tropical scientists believe that specialized insects and pathogens that primarily attack a single tree species check that species from dominating the forest. Foresters in both the temperate and boreal forests know that on plantations of trees or in vast stands of few species, such as the white or black spruce trees that blanket the upper latitudes, pest outbreaks are common. Such eruptions are even more common in tropical tree farms where trees of the same species are planted in large numbers—essentially a monoculture. Besides, in tropical regions, plant herbivory can be more intense than in forests closer to the poles; in the latter, the long winters suppress the activity of pest species. In contrast, tropical plant eaters are typically active every day of the year, especially in the wettest rain forests. If a given tree became common as a seedling, sapling, or small tree, the plant predators to which that species was susceptible would proliferate in the presence of more abundant food and would soon devour the increased supply. The combined effect of these leaf chewers, suckers, and defoliators is that individual trees that are spaced far from other members of their species—often a distance of a hectare or more—have higher survival rates. So the very nature of the hot, humid, and stable climate, especially warm winters, that leads to considerable diversity of trees also fosters a rich coterie of the creatures that inhibit their spread.

An allied theory accepted by many tropical biologists has held that patterns of seed predation and dispersal best explain the low density of canopy tree species. According to this hypothesis, if most of the seeds of an adult tree fell beneath its crown, waves of seed predators, from ants to agoutis, would finish them all off. Only those seeds carried a safe distance from the parent by some animal, be it a fruit-eating bird, mammal, or even fish, would find a better and safer germination site. Biologists note that as much as 90 percent of all plants in tropical forests produce fruits that are dispersed by animals, whereas in the temperate zone the percentage is much lower, often as low as 10 percent, and most tree seeds are dispersed by wind, water, or gravity. Left relatively undisturbed and undiscovered by the seed predators congregating at the base of the parent, a distant seed would germinate and prosper. This phenomenon also contributes to tropical canopy trees growing far apart from one another.

For George and Sue, knowing the distribution of the trees that were important resources for the macaws and monkeys was a vital research question. So, like many before them, they worked with teams of botanists to map the distributions. There was one simple but pervasive problem, though: how to identify all those trees in the first place. Until the late 1980s, a lack of reliable identification guides to nonflowering and nonfruiting plants had held back progress in field tropical biology. Botanists relied on flowers and fruits to identify species, but few plants were ever in a reproductive phase at the same time, and the leaves of many species all seemed to converge on the same basic shape—narrow, without teeth on the margins, and with an elongated tip. Not much to go on.

Then came Alwyn Gentry, a curator at large with the Missouri Botanical Garden. While in the field he noticed subtle features that escaped the attention of most collectors and curators, many of whom had prematurely decided that leaves alone, or other vegetative features, would be of little value in on-the-spot species identification. Gentry worked with another fine tropical botanist, Robin Foster, to identify tropical trees and shrubs, including many rare ones, relying only on parts on display every day—leaves, stipules, glands, twigs, bark, thorns, trunk, or exposed roots. The clues to identifying species from these features never appeared in a botanical text. That comprehensive field guide lived in Gentry's head.

"Fortunately, he started to write it all down for the rest of us," said Adrian Forsyth, who helped finance the publication of Gentry's magnum opus, *A Field Guide to the Families and Genera of Woody Plants of Northwest South America (Colombia, Ecuador, Peru).* While the book was in its final editing stage in 1993, Gentry's small plane crashed into a mountainside in western Ecuador. Gentry died at the age of forty-eight, approaching the apex of his career.

Raul Tupayachi is a top Peruvian botanist and the head of George's botany field team. Raul's uncle had worked for Gentry and handed down his inside knowledge. On the second day of fieldwork, George, Raul, and I hiked on a trail through the Tambopata forest to join the plant collectors. En route, we came upon a carpet of beautiful white blossoms resembling pincushions. I stopped to pick up some of the fragrant *Inga* flowers that had fallen from a tall tree. "You know, Eric," Raul said, "we've already recorded twenty-two species of *Inga*, and we may have twenty-five. It's the most diverse group of trees in this part of the Amazon." This was a striking instance of the interplay of rarity and abundance in the tropics.

What's more, many of those *Inga* species could be found on the same hectare. According to standard textbook niche theory, the reason so many tree species, and so many within the same genus, can share the same hectare of forest is that the trees divide up the available niches in microvariations of their environment. These variations occur among many different dimensions, such as light levels; concentrations of nitrogen, iron, or potassium in the soil; and relatively wet or dry spots on the forest floor. Niche theory holds that the ecological separation of many species is a reflection of their differing abilities to use such limiting resources; scientists merely lack the tools to detect all those dimensions just yet. However, a visit I had made to a tropical forest research site in the Lambir Hills of Sarawak, Malaysia, a few years earlier made me wonder if at least one part of niche theory deserved a recall. There in the northern Borneo rain forests, as many as twenty-five species in the genera *Shorea*, *Parashorea*, and *Eugenia* all occurred on the same hectare, like the many species of *Inga* in Amazonian Peru! On what axis of limiting resources required by plants—nutrients, light, moisture—could trees micromanage their needs in order to live so packed together?

The Lambir Hills site was part of a remarkable experiment to study and map rain forests at an unprecedented scale—fifty hectares, about the size of a small vineyard. The first to do this were ecologist Steve Hubbell and botanist savant Robin Foster. Their landmark study on Barro Colorado Island, Panama, in the early 1980s illustrated the patterns seen everywhere in the tropics: an incredible diversity of tree species—cresting at 125 to 150 species per hectare—with virtually all appearing on their plot at very low densities. Among some species, an adult tree was found only every few hectares. How different this is from a temperate zone forest, where on a single hectare one might find the same number of individual stems but fewer than two dozen tree species. Since then, twenty-three plots of similar size have been intensively studied across equatorial forests from Peru to the Philippines. Each has yielded census results similar to those from Hubbell and Foster's fifty-hectare samples.

Reanalysis of some of these large data plots yielded, in the first decade of this century, yet another paradigm of tree diversity and local rarity. Steve Hubbell proposed a theory of "functional equivalence" in which he suggested that many tropical tree species function as ecological duplicates of one another and the species that grew in a given location in the forest was determined purely by chance. Thus, many plant species exist in tropical forests because they all have an equal chance of occupying spaces that open up when trees fall or die. Rarity in tropical trees and vines is thus in part a function of limited habitat required for successful propagation of a new generation, in particular a shortage of space for seedling survival, such as the infrequent light gaps, or openings, in the forest for species whose seedlings and saplings require intense sunlight. A tree that is the only one of its kind in a forest stand could also be a recent immigrant whose seed somehow managed to disperse and survive far from its clan. The take-home message of Hubbell's work was that the forest was not in a permanent state of balance. The commoners might fluctuate in number, but species that were extremely rare—say, one individual per fifty hectares—could also just as easily disappear.

Hubbell may be correct in saying that trees so rare that only a single individual is found on average per fifty hectares could easily disappear. For example, tropical ecologists have discovered that the best germination sites for each species, such as open areas created by treefalls that are quickly shaded out, may be short-lived or unpredictable. Consequently, few of a tree's seeds may germinate or survive past the seedling stage. Yet trees that have evolved to live hundreds of years and produce 10,000 seeds per year have their own built-in protection for persistence. The production of viable offspring can afford to wait for the favorable but infrequent germination conditions that may depend on a fire, windstorm, or other event. After all, to be successful—to pass on its genes to the next generation—an individual need only replace itself once in a hundred years, and canopy trees live to be much older than that. The life history of a rare tropical canopy tree is a study in patience.

These hypotheses proposed to explain patterns of rarity in tropical trees are not mutually exclusive, and all are likely to be part of the answer. And, most interestingly, none is designed to explain how the large number of species was generated in the first place. That piece of the puzzle remains unsolved.

~

On a tranquil morning seven days after arriving in Madre de Dios, George, Sue, and I started back up the Tambopata. A heavy mist on the river kept the kingfishers grounded, but the sound of the outboard engine stirred awake slumbering capybaras on the floodplain islands as we approached. The world's largest rodents had found a safe haven from jaguars in the middle of the channels. There was not much green on the sand and stone beaches for their breakfast, but there were no big cats or anacondas, either.

After a night's rest, we continued our journey upriver before dawn, making our way to a blind created by some washed-up trees on a floodplain island. Across the channel an exposed clay cliff, about ten meters high, appeared before us like a vast canvas painted in a dull ocher wash, lifeless. "Just wait," Sue whispered. "You won't believe this."

With the first warming rays of the sun, flashes of blue, yellow, and gold flitted along the cliff face, followed by red and green with dashes of scarlet, blue, and chestnut. The *collpa*, or mineral lick, was alive with macaws and parrots, squawking and chewing on the mineral-rich earth. Not a handful of birds but hundreds of them represented five species of macaws—scarlet, red-and-green, blue-and-yellow, chestnut-fronted, and red-bellied—and at least nine species of parrots. We counted blue-headed, mealy, yellow-crowned, orange-cheeked, and white-bellied parrots, as well as dusky-headed, cobalt-winged, and white-eyed parakeets and, rarest of all, an albino blue-headed parrot. This must have been one of the densest concentrations of natural color anywhere. The macaws and parrots are believed to eat the earth to obtain sodium—an element in short supply in the food they normally ingest—and perhaps to help detoxify some of the chemically laced seeds they consume. There may also be social functions to this gathering. Whatever the causes, in this Kingdom of Rarities, the rare aggregation of multi-hued birds dazzled us—as did the swarms of brightly colored butterflies fluttering along the shoreline as we returned to the boat.

A cold front, known locally as a *friaje*, had moved upriver with us, and we spent the next two days in a chilly rain forest—who would imagine 10°C (50°F) in the Peruvian Amazon? The Madre de Dios region seemed caught in a collective shiver, and at night the temperature dropped precipitously. The effects of the friaje were most evident in the absence of local nocturnal wildlife seen or heard. We spent one night in a blind above a mineral lick usually frequented by salt-seeking tapirs. None appeared. Bats were scarce along the trails they typically used to commute between their roosts and their feeding areas. The silence was also conspicuous. Quiet

were the lead vocalists, the kinkajous—rain forest relatives of raccoons—the night monkeys, and the amphibians and insects that
provide the background chorus.

That the cold weather disrupted the normal rhythms of the forest dwellers could explain what we saw late the next afternoon. We
were heading back to camp when our sharp-eyed boatman pointed
to a sandy stretch along the Tambopata floodplain. He slowed the
boat and whispered, "*Mira!* Jaguar!"

Reclining on the beach was a large male jaguar. He was magnificent and quite unperturbed by the close approach of the boat.
The driver turned the prow upstream, and we sat and watched the
stationary cat. Ten minutes went by, then twenty. Perhaps the sand-
bank held the warmth of the sun's rays in the approaching dusk,
and after a frigid week in the forest the jaguar was content to rest
and warm his bones. We were so enraptured by our sighting that
we ignored a smaller boat idling nearby until we noticed the three
men in the craft eyeing the same splendid cat. Tucked below the
gunwale was a barely concealed rifle. To foil the hunters, we waited
until it grew too dark for them to see the outline of the jaguar.
With no chance for a kill, the men headed back downriver. We
disembarked and walked toward the jaguar, urging it back into the
forest for its own protection. In their years in Peru, George and Sue
had never seen such a calm jaguar, much less have to save one from
other humans.

That evening, the unforgettable encounter with both predator
and poacher was at the center of our conversations. "George, if the
poachers had arrived before we did, there would have been one less
jaguar hunting along the floodplain of the Tambopata. But take
it to the extreme. What if protection broke down completely and
hunters shot every last jaguar in these parts? How would this ecosystem be different?"

George didn't hesitate. "You remember John Terborgh's studies
in Venezuela at Lago Guri, don't you?" George was referring to one
of conservation biology's leading field men, a hero to us. Terborgh's

innovative natural experiment in Venezuela involved censusing the inhabitants of islands created by dam impoundments in once continuous forest. Some of the human-made islands were too small to support even a single jaguar or other felid. Terborgh showed that when such top predators are lost in a system, the forest changes rapidly because there is a swift increase in the numbers of sloths, monkeys, deer, and peccaries that eat plant matter and all the midsized vertebrates that eat insects. The forest's low vegetation becomes less dense because populations of species such as leaf-cutting ants explode; while the howler monkeys defoliate the tree canopy, the ants clear out the understory. The absence of large carnivores creates a different kind of kingdom, minus the usual ecological checks and balances, and allows, at least for a short time, explosive growth of certain herbivores—mammalian and insect.

The impact that rare top predators can have on ecosystems was made even clearer in 2011. In a review paper in the journal *Science*, James Estes, John Terborgh, and their colleagues introduced a new phrase into the ecological lexicon: the "trophic downgrading of planet Earth." On land, in streams and lakes, and in the sea, the removal of apex predators—jaguars, pumas, salmon, sea otters, sharks—has had a profound and cascading effect, termed "trophic cascades," on the species layered under them in the food pyramid. The Terborgh study in Lago Guri mentioned earlier was but one example of a trophic cascade that had many parallels in other biological realms. The loss of sea otters, for example, had a profound effect on the distribution of kelp forests as the urchins that grazed the kelp were no longer kept in check by the otters. All of the species that depended on the dense kelp forests for food and habitat were affected. In freshwater systems, the drastic reduction of native salmon runs has reduced the amount of enrichment that occurs upstream as salmon spawn and die and their decomposition releases vital nutrients into the system. Other fish also become more prevalent in the absence of this predator. On land, removing the top predators means that the effects trickle down in the ecosystem, reducing the productivity of the soils, the diversity of plants, and the amount of carbon sequestered in forests.

Terborgh's long-term study, begun in 1990, of what had become virtual island ecosystems ended around 2005 with a surprising outcome. It supported one of the signature theories about how the natural world works: that the tropical world is green only because leaf-eating creatures, from howler monkeys to leaf-cutter ants to insect larvae, are kept in check by their predators. At the same time, it unexpectedly undermined an opposing argument, that the nasty chemicals in the leaves were enough to keep herbivores at bay. This top-down effect large predators exert on the ecosystem stands opposite the bottom-up effects we saw in the operation of the invertebrate leaf and seed eaters in regulating the distribution of tropical trees. Both may be at work at the same time.

The role of peccaries is also important in shaping the Amazon rain forest in this region. Peccaries consume fruit and seeds from more than 400 plant species, far more than any other fruit eaters in the forest. They likely disperse the seeds of about 250 of the ingested species but act as influential seed predators for the other 150 species. They also trample many seedlings and suppress their recruitment. But for peccaries to have such a landscaping effect on the forest they need to be abundant, not nearly hunted out and rare. On the other hand, should jaguars and pumas disappear, the interactions between peccaries and plants would increase with their populations and change the composition of the forest. Harald Beck, a peccary biologist and Terborgh protégé, found that in areas where peccaries no longer roamed freely because they were hunted out, there was a dramatic increase in seedling dispersal of a common palm genus, *Iriartea*, whose seeds peccaries destroy in consumption. Peruvian forests without peccaries and jaguars look different from forests where they have been present, in part because palms are less common in the understory when those species are present. Peccaries, along with rhinos elsewhere, certain other large mammals, and some birds, are able to serve as ecosystem engineers. As such they play vital ecological roles, often disproportionate to their numbers or biomass, in shaping the structure and species makeup of the rain forests. These are keystone species: their demise often triggers a collapse or a dramatic change in that system, much as the keystone in an arch is an essential piece in the design of that structure. The loss of a keystone species often triggers a trophic cascade.

Of course, one doesn't have to purchase a ticket to Puerto Maldonado, or any tropical wildland, to see the effects of such a phenomenon. A walk through a deciduous forest in the eastern United States illustrates the ecological damage of an ongoing trophic cascade. The eradication of pumas and wolves from these forests, along with their fragmentation, has led to a proliferation of white-tailed deer. Absent their predators, deer have become a nuisance to motorists and gardeners and a health threat, serving as hosts for the ticks that spread Lyme disease. The change has not been good for native birds such as wood warblers, either. The understory vegetation in many eastern forests has been greatly reduced or even eradicated by the hungry hoofed mammals. As a result, the hooded warbler, worm-eating warbler, Kentucky warbler, and Canada warbler, among other species that nest on or near the ground, have declined because overbrowsing by deer exposes their nests and nestlings to predators.

In the rain forest, monkeys also serve as a great example of keystone species in the canopy. I mentioned to Sue one day as we were scanning the canopy for primates, "Your saki monkeys are probably safe from hunters because of their cryptic behavior . . . But what about spider monkeys? Don't hunters consider them a delicacy? Do we know what happens when a forest loses its fruit-eating monkeys?"

"Spider monkeys are often the first to be hunted out in most rain forests," Sue responded. "It seems animals that eat mostly fruit— spider monkeys, macaws, parrots—all have tasty meat the hunters go for." Then Sue mentioned the work of a World Wildlife Fund postdoctoral student, Gabriela Nuñez-Iturri, who compared seedling presence and abundance in areas where spider monkeys and other fruit eaters are still common and protected with that in areas where these species have disappeared. At sites where primates had been hunted for thirty to forty years, seedlings and small juvenile trees whose seeds are dispersed by the larger primates were reduced by nearly 50 percent, whereas the seedlings of wind- and gravity-dispersed plants became 284 percent more common. Her study showed that the composition of seedling and small juvenile tree species that ultimately regenerate future forests differ markedly in hunted forests as compared with protected forests where primates and large fruit-eating birds still thrive.

The loss of the big, rare tropical trees from logging would also have a dramatic ecological effect. By combining Greg Asner's laser-scanning data on the carbon density of the rain forest with the information Sue, George, and Gabriela had gathered, we have recently learned even more about how the fates of jaguars, monkeys, macaws, and peccaries and the future of tropical forests are closely linked. The tree species that hold the most carbon in the forest are the ones with the densest wood and with seeds that are dispersed by birds and mammals, and they are typically rare per hectare. If those birds and mammals disappear as a result of hunting, the rare trees over time are likely to be replaced by an aggregation of trees that have much lower carbon densities and much lower diversity. If soaking up carbon dioxide is a global benefit conferred by tropical rain forests, then monkeys and large fruit-eating birds, by the seeds they disperse, help to create the most carbon-dense forests. We can thus see how the loss of primates and birds from a rain forest not only removes a keystone species but also contributes to a trophic cascade, a process that typically leads to a more impoverished, less stable environment.

If monkeys, macaws, and jaguars are so ecologically valuable, how can we make them worth more alive than dead to the local people—those who live in or at the edge of the rain forest and rely on its bounty? George has been working closely with Adrian

Forsyth and their Peruvian colleagues to design a network of parks and conservation areas to keep this Amazonian region as intact as possible. If jaguars need a lot of space and the largest reserves are still too small to hold what biologists consider a genetically healthy population, linking reserves through corridors jaguars use can greatly enhance the conservation effect. Maintaining connectivity between reserves increases movement and ultimately gene flow. In the Amazon, such landscape-scale conservation with respect to large predators and other roamers is still possible with the proper incentives. One promising program, called the wildlife premium mechanism, would pay poor Peruvians who collect Brazil nuts for a living to become stewards of the forests and protectors of jaguars and pumas. By agreeing to maintain the dispersal corridors for jaguars among Manú National Park, Tambopata National Reserve, and nearby Alto Purús National Park, the communities could help secure a landscape of several million hectares and receive financial benefits. The plan, which is still in its infancy, may hold a key to a future with jaguars and pumas and the best inoculation to a cascading decline of Amazonian diversity beyond its treasured parks and reserves, where jaguars still venture out onto the beach.

Chapter 4

The Firebird Suite

*A*N UNBROKEN VISTA OF STUNTED JACK PINE trees dotted the northern Michigan landscape. Maybe our guide had mistakenly led us into a Christmas tree farm. The only sizable trees in view were a few skeletons—blackened snags that remained after a blaze a few years back. This unremarkable setting seemed like the last place one would search for rarity.

"Keep watching the snags," advised our escort. "The males prefer them as singing perches." A few small songbirds flitted about the pine grove and our group of birders snapped to attention. False alarm. The distant silhouettes became sparrows and juncos—perfectly nice birds but not the ones we were here to see.

Then, almost simultaneously, three male Kirtland's warblers rose out of the dense underbrush and perched, one per snag, to advertise their individual glory. A nearby singer threw back his head and uttered the anthem of the jack pine, a loud and lively *"Flip lip lip-*

lip-tip-tipCHIDIP!" Within seconds, a fortune's worth of spotting scopes, massive telephoto lenses, and binoculars were trained on the handsome male. The fifteen-centimeter-long bird sported a bright yellow breast streaked in black; a dark mask over his face, highlighted by white eye rings; and a slate-blue back—a perfect marriage of classic understated plumage and a splash of color. We could hear more singing males in the distance, all staking out their territories before the females arrived. "This population is rebounding rapidly," the guide was telling us, "and the current estimates for this year are about 1,770 singing males in the wild."

Gleeful smiles signified our good fortune on that May day in 2010. For almost everyone in the group, this was a first-ever sighting, a "life bird." Some had traveled across oceans to record the event. Secondary for them was the news that this stunning bird was staging a comeback. But that was the purpose of my visit—to understand better how rare species with particular habitat preferences manage to persist and even recover when perched on the edge of extinction.

If a shrine to rarity exists, the jack pine woods near Grayling, Michigan, is nature's Lourdes. Each spring, thousands of birders journey to this small town in the state's Lower Peninsula, about five hours' drive north from Detroit, to encounter the rarest breeding songbird in North America. Territorial males return in early May, having winged in from their wintering grounds in the Bahamas. The females linger on the islands a bit longer, fattening on fruit and insects, before beginning the perilous 2,500-kilometer migration back to their only significant breeding site. From the exotic Caribbean islands to the stunted jack pine woods on the outskirts of Grayling, such is the seasonal arc of a most uncommon species.

The day after our birding excursion, a group of young marchers fidgeted under the bright morning sun, eager to start the celebration. It was the third Saturday in May, time for the annual Kirtland's Warbler Wildlife Festival. A roving reporter asked a young girl in line, "Do you know why the Kirtland's warbler is so rare?" In response, she could have pinpointed part of the answer by quoting a verse from Dr. Seuss, the one they teach in science class about the Nutches, "who live in small caves, known as Niches." Unfortunately for the Nutches, there simply weren't enough niches to go around.

The singing male Kirtland's we witnessed in the pine grove had to compete for his niche, limited breeding space. Kirtland's warblers have historically nested only in stands of small-stature jack pine maintained by natural fires. This extreme preference about where they will or will not make their nest gives them, and a large number of other species, a special place in the Kingdom of Rarities. There are species that have limited ranges, such as the golden-fronted bowerbirds, and those that live at low densities, such as the jaguars. The Kirtland's warbler, singing boldly from its perch in a fire-prone stand of young jack pines, meets those conditions and raises the ante. Beyond having a narrow range and low numbers, it represents another dimension of rarity: extreme habitat specialization. By this I mean a species that breeds only in a particular habitat, feeds only there, or requires some other feature there to thrive. That could be a cool, dry, or wet sanctuary to maintain an ideal body temperature or the proximity of a river or stream from which to drink several times per day.

In the grand scheme of nature, many species are quite particular in their habitats. Arctic lemmings and other species that reside in the tundra could not live anywhere else, for example. Parrot fish are literally a fish out of water beyond their preferred tropical coral reefs. Even the ubiquitous prairie dog could not survive anywhere except a grassland. All these habitat specialists are, fortunately, adapted to living in an ecosystem that is widespread in the world, at least today. But others, including the Kirtland's warbler, are specialists on highly restricted habitats. I came to see how such extreme specialists among species persist and recover, but I also wanted to explore in what circumstances such a narrow habitat bandwidth becomes a liability. All of the patterns now present could be in flux: as climate change scientists are telling us, the distributions of preferred habitats may change dramatically over the coming century. Those species that are specialists on widespread habitats and are common today may find a vastly shrinking range tomorrow. The narrow-range Kirtland's warbler may thus be a bellwether species to help us better understand how species cope in a more restricted area.

Extreme habitat fidelity: is it a cause of rarity itself, or a condition of it? Actually, it can be either or both, depending on the circumstances. The linked questions have been central to the debate on what lies behind patterns of rarity and abundance in nature. Our understanding of extreme habitat fidelity as a condition of rarity owes much to the work of the population biologist Deborah Rabinowitz. In 1981, she developed a way to think about rarity that remains influential today: to a species' range size and its population density, she added a third quality—its loyalty to a particular habitat—as a condition of rarity. One thing is clear: if highly selective breeding habitat alone explained a major piece of the rarity puzzle, Dr. Seuss's classic could replace bookshelves of ecology texts. But there is much more to the story. For example, if today you must go to Grayling to see Kirtland's warblers with ease and in appreciable numbers on their breeding grounds, time travel back before the last ice age might well have offered more viewing options. Some scientists believe that this species was probably more common prior to the last glacial period, when the jack pine forests were more widespread. Grayling sits on the edge of the bird's historical range. Some biologists predict that under several climate change scenarios projected for northern Michigan, the future for this species could be greatly curtailed, but that still remains conjecture.

At its low point, in 1971, the global population of Kirtland's warblers dropped to around 400 birds. At that time, only about eighteen square kilometers of suitable immature jack pine habitat remained, and that was in Michigan. The warbler joined the first US endangered species list in March 1967 as one of seventy-eight native animal and plant species then under threat of extinction, and it remains America's most imperiled breeding songbird.

How frequent is the condition of extreme habitat specialization in nature, and how might interventions on behalf of the Kirtland's warbler inform efforts to save species with similar distributions? Viewed through our magnifying lens of rarity, maybe the seemingly boring Christmas tree farm held a lot more interest after all.

Standing in the midst of the jack pine forest on that May day, I still couldn't get over the sameness of it all. Having lived and worked in species-rich tropical rain forests, I have a natural aversion to monocultures. If the view from the spotting scope brought rarity into sharp focus, the naked eye swept a panorama that seemed like a biological wasteland. Why did the Kirtland's insist on nesting *here*? If I pretended to be a northern Seuss, I might ponder:

What's this business about jack pines, do tell!
Why so picky where you pick to dwell?
And why nest on the ground
where skunks abound,
and heavy rains cause your chicks to drown?
Mrs. Kirtland, why not raise them in town?

And what was the fate of those birds that failed to find a home in a young jack pine grove? Did their offspring survive at all? Skunks and other midsized nest predators do eat Kirtland's chicks. But did chicks survive better when such top predators as wolves, bears, and pumas kept the population of skunks, raccoons, and their egg-snatching ilk in check? My field notebook held more questions than answers.

The singing male above me kept at it. In the heart of his territory were huckleberry bushes, aromatic sumacs, and sweet ferns (a shrub in the bayberry family), but the tree layer was virtually a pure stand of young jack pines. The pines were hunched and distorted, partly because they were growing on Grayling sands, an acidic, porous, nutrient-poor soil. The trees were about eight years old, perfect for this warbler, which builds its cuplike nest on the ground under

*Male Kirtland's warbler (*Dendroica kirtlandii*) on a jack pine branch*

the spreading pine boughs. Saplings younger than five years of age offer poor cover, while the upper branches of jack pines older than twenty years block the sun from reaching the ground. At that age, too, the lower branches of older trees drop off, and overhanging grasses become shaded out. This exposes the ground nests to predators, which besides skunks include snakes, thirteen-lined ground squirrels, and blue jays. Kirtland's are also particular about substrate, the surface on which an organism grows or is attached. Rainwater percolates quickly through Grayling sands, so shielded nests and the surface soil layer stay dry even after summer cloudbursts. Nestlings left sitting in puddles or on waterlogged soils suffer heavy mortality.

The bold male in the snag spent the next fifteen minutes throwing back his head, bobbing his tail, and filling the air of the jack pine grove with his lively song. Such lovely plumage in the bright sunshine brought cheer to everyone. Nearby, Nashville warblers began singing. Kirtland's and Nashville warblers, along with the other fifty-four species of New World warblers that breed in the United States and Canada, are part of nature's balancing mechanism, as these dedicated caterpillar eaters, when gathered in numbers, clean the forests of larval insects harmful to trees. They are also among the most stunning of pest removers. Natural selection, using delicate brushstrokes and a colorful palette of yellow, green, chestnut, orange, red, blue, and black, has fashioned an exquisite array of tiny creatures—only one warbler species is larger than a sparrow.

Unlike most other members of the warbler family, the Kirtland's was discovered relatively late in the history of North American ornithology. It was first described by taxonomists in 1851, when a male was collected on the outskirts of Cleveland, Ohio. The songbird was named in honor of Jared P. Kirtland, an Ohio physician, teacher, horticulturist, and naturalist who assembled the first lists of vertebrates for the Buckeye State. Ironically, by the time the Kirtland's warbler was named, it was already experiencing a range collapse, and by the time of Kirtland's death, in 1877, it may have disappeared entirely from Ohio.

The bird's wintering sites first became known in 1879, when a specimen was collected on Andros Island in the Bahamas. Since then, wintering Kirtland's warblers have been found on other islands in the Bahamian archipelago, on the Turks and Caicos Islands, and on Hispaniola. Yet only in 1902 did biologist Norman Wood find the first nest, in Michigan's Oscoda County. Like many of the more than 200 species of migratory songbirds that nest in the United States and Canada, the Kirtland's warbler actually spends eight months of the year on its wintering grounds.

Among the birders in our makeshift Grayling caravan was Sarah Rockwell, a University of Maryland doctoral student who in 2006 had chosen this bird to be the center of her life for the next five years. When I first heard of Sarah and her work through a colleague, I wondered why in the world she would undertake such a study. Successful PhD dissertations in biology are all about large sample sizes. Piles of data are essential for the rigorous statistical analyses and testing of hypotheses that graduate committees demand from students. Indeed, professors typically divert their more idealistic or naive students from the futility of field studies on rare species.

A major reason she became interested in studying Kirtland's warblers, Sarah explained, is that she really liked the idea of contributing to conservation efforts in her home state. When working with an endangered species, any data one can gather on its ecology might be immediately useful to managers. "Besides," she noted, "I actually have no problem with sample size. Kirtland's are quite abundant in the right habitat patches. I have found up to one male per hectare in the busiest locations."

Sarah was not the first to study the Kirtland's for her doctorate. In the early 1990s, Carol Bocetti compared the breeding success of Kirtland's warblers nesting in managed (that is, harvested) jack pine plantations that had been started a decade earlier with the breeding success of Kirtland's nesting in jack pine groves that had naturally regenerated following wildfires. Remarkably, Carol found that the density of territorial males was similar in managed and natural stands.

Carol's work and that of others illuminated the basics of the warbler's summer range and breeding biology. Meanwhile, in the Caribbean, Joe Wunderle, a noted tropical ornithologist, and his research team had started to put together a picture of the winter ecology of the species. Sarah saw her opening when she heard of Joe's study. "Are there aspects of their winter life," she asked—the climate where they stay, their habitat use, their diet—"that affect reproductive success and survival?"

Sarah had predicted that birds that arrive early on the breeding grounds would be the most successful parents because they would occupy the best sites during the long winter stay in the Bahamas. For a long time, biologists were unable to track individual birds throughout the year or match wintering conditions to breeding success up north. Some who studied neotropical migratory birds simply assumed that these species led separate lives in the two locales and that their winter behavior had no effect on their summer activity. Sarah's study questions, supported by collaboration with Joe's Caribbean-based team, required her to follow the fate of her birds through an entire annual cycle.

Many dangers confront these long-distance night travelers, about 60 percent of which die in their first year, Sarah noted. Along their 2,500-kilometer route, they must avoid collisions with radio towers, transmission lines, windows of homes and office buildings, and other obstacles. En route and on the home breeding grounds, migratory songbirds must avoid being eaten by predators. The fate of migratory Kirtland's was no different from that of many species of migratory songbirds. But for extreme rarities with breeding numbers as low as theirs, the loss of many adults could be devastating.

Joe had his own set of unanswered questions that meshed with Sarah's. The first addressed what winter habitat qualities the birds preferred. Over several years, his team had identified the conditions the warblers favored in late winter just prior to migration, which might have a bearing on their rarity and their survival. On Eleuthera Island, where he found a small concentration of wintering Kirtland's, the birds eat a lot of fruit and prefer snowberries, black torch berries, and the fruits of wild sage. That means the birds hang out in second-growth areas where these plants occur. Habitats resulting from second growth eventually diminish in size, however, unless they are preserved by some big disturbance event. Historically, disturbance meant fires or hurricanes. Today, however, it's human activities. "So we hypothesized that disturbance by humans actually favors the spread of fruit plants used by the warbler."

The test of their hypothesis led to one of the most surprising and counterintuitive discoveries in the history of ecology. "In the

Bahamas," Joe told me, "at least for Kirtland's warblers, goats are kind of a blessing!" The ecologist's scourge of the earth, the common goat, was resurrected as a hero. Here's why. The goats ignore the leaves and stems of plants that supply the Kirtland's with fruit and instead consume the competing plants. Thus they delay plant succession, the process by which a natural community moves from a simple level of organization to a more complex assemblage. Succession is a natural process that typically occurs after some form of disturbance has simplified the system. "In fact," Joe continued, "our work shows that goat farms make great Kirtland's habitat." There are probably dozens of biologists who work in Hawaii, the Galápagos Islands, New Zealand, and other islands where goats were never part of the native mammalian fauna who would pay Joe handsomely to keep quiet. Goats are often the worst culprits in the extinction of rare native plants, and they typically outcompete native herbivores for forage. On Eleuthera, at least for Kirtland's warblers, they were a force for good, especially on parcels where farmers rotated their goat herds among fenced rangelands to prevent overbrowsing.

Many questions remain unanswered about the Bahamian winter getaway. For example, as in Grayling, natural fire regimes in the Bahamas may provide the disturbance essential to support the fruiting shrubs upon which Kirtland's warblers depend. But are today's managed fire regimes too different from wildfires of the evolutionary history of the Bahamas? Are goat farms and smart, "green" bulldozing practices enough to offset the suppression of fire? Finally, little is known about the habitat needs of the warblers during their critical three-week migration from Eleuthera to Grayling.

⌒

As we walked through the pine grove in Grayling, Sarah reflected on my first impressions of jack pine landscapes. "I had the same response as you did when I first arrived!" she exclaimed with a laugh. "It may not be very spectacular at first glance, but the more time you spend in the jack pine forest, the more you appreciate that it has its own beauty. You can be on a little hill in the early morning, watching the mist that settles in the low places when the sun rises. I often see spruce grouse, upland sandpipers, porcupines, fawns, occasionally a coyote. There really is a lot to like about it!" Sarah was starting to think like her study bird.

Her biggest challenge had been learning to nest-search. The difficulty of finding nests of a study species, or the inability to do so, has probably derailed more potential PhD dissertations in ornithology than any other obstacle. Sarah stopped to look under a pine. "The nests are so well hidden that you can't find them through systematic searching or looking on the ground in a known territory. You have to use parental behavior as your key."

First, she had to spot a female carrying nest material or a male carrying food and follow the bird back to the nest. The parents, though, saw her as a predator and didn't want her to find their vulnerable eggs or young. "If you're not sneaky," Sarah said, "they'll never go to the nest; they'll just fly around in circles with the food." She learned to hide behind jack pines and crawl on her stomach to see under the branches and, eventually, watch parents go to their nest.

As she grew more adept at keeping a low profile, she picked up other tricks. For one, the males tend to sing while they deliver food, but the quality of the song is different, more muffled and gurgly. The closer they get to the nest, the quieter and sweeter the song becomes. "Then the male goes silent while feeding the incubating female or nestlings, and after a few seconds he'll pop back up high in a jack pine, singing his regular, loud-and-proud song. You can almost locate a nest by ear alone." Of course, it takes experience. In her first year, Sarah and her team found sixty-three nests. The next year, they almost doubled that total.

To learn about the effects of overwintering and summer survival, Sarah and her colleagues net as many birds as they can, band and measure them, and collect blood samples. This provides data on body condition, and she uses the blood for stable-isotope analysis to help infer winter diet and habitat quality.

Stable-isotope analysis, an exciting new approach in ecology, involves measuring the ratio of the heavy form of an element to its lighter form. For instance, carbon atoms with an atomic mass of 12 are the most common in nature, but a small proportion have an atomic mass of 13 because of an extra neutron. "What's really interesting is that isotopes of some elements vary in a regular way that is useful to ecologists," Sarah explained. "For example, the ratio of ^{13}C [carbon 13] to ^{12}C [carbon 12] in plants is greater in dry habitats than in wetter environments—we call this being more enriched in $\delta^{13}C$" (a measure of the ratio of stable isostopes carbon 13 to carbon 12). Habitat quality for many overwintering migrants varies along this same wet-to-dry gradient that can be measured by $\delta^{13}C$. A habitat-specific stable-carbon ratio is incorporated into birds' tissues through the fruits and insects they eat during the winter. Nitrogen is another useful element; the ratio of ^{15}N (nitrogen 15) to ^{14}N (nitrogen 14) varies by trophic level, with organisms further up the food chain becoming more enriched in $\delta^{15}N$ (the measure of the ratio between heavy and lighter nitrogen). So Kirtland's warblers that consume a greater proportion of insects than fruit will have a higher $\delta^{15}N$ signature in their tissues.

When Sarah captures a bird soon after its arrival in Michigan and takes a blood sample, its red blood cells retain the stable-carbon and stable-nitrogen signatures of its late winter habitat type and diet. Researchers have called isotopes in birds "flying fingerprints." The real advantage of this technique is that Sarah doesn't need to know exactly where each of her birds has spent the winter and what they have eaten. She can obtain the critical information from the stable-isotope fingerprints that they've carried with them to Michigan.

Ultimately, the critical data Sarah needed were the standard metrics of animal biology: reproductive success. In this case, it was how many young each pair of Kirtland's warblers raised to fledge. Sarah and her team spent June and July searching for nests. Each time they found one, they recorded a GPS location; later, they returned to the site to count the number of nestlings present. Sometimes this scrutiny didn't go over well with Mrs. Kirtland. If flushed off their nests, the angry mother birds would perch right next to the researchers and practically peck their hands while they counted eggs. After fledging should have occurred, a researcher returned to the nesting territory to confirm it by looking and listening for fledglings, angry chirping parents, or parents carrying food after the nest is empty. Not all is Disneyesque. Sarah reported that about 25 percent of the nests to which the researchers returned had been depredated by egg and nestling eaters.

A warbler flew up to its singing perch. I asked Sarah the question the dissertation committee at the University of Maryland would likely put to her: "So what have you found so far, and how will it help managers?" Sarah's isotope analyses gave her the answers. After five field seasons, her preliminary results showed that Kirtland's warblers originating in wetter winter habitats, as indicated by more negative carbon-isotope ratios, arrive first, although the pattern varies by year. The early birds have first pick of the best territories and mates and have more time to breed. This additional time on the breeding ground offers them a chance to renest if predation occurs or even to attempt a second nest if time permits. Also, hatching failure and predation tend to be less earlier in the season. "The bottom line, I've found, is that the arrival date of males is significantly correlated to the greater number of fledglings they raise that year," Sarah said. She also found that following wetter March weather in the Bahamas—the month prior to spring departure—Kirtland's warblers tend to arrive sooner in Michigan, be in better body condition, initiate nests sooner, and fledge more offspring.

Considerable research has linked winter rainfall, food supply, and overwinter condition of other migratory birds. Now Sarah was finding that these factors carry over to affect birds in the subsequent breeding season as well. She hoped her findings would help managers choose the best-quality habitats in the Bahamas to conserve—those that retain moisture better through the winter dry season. This factor would be especially important if long-term drying trends in the Caribbean due to climate change occur as predicted. For the first time, biologists realized that to understand how the population size of Kirtland's warblers is regulated and what the species' prospects are, they need to understand what the limiting factors are—such as winter climate and habitat—at each stage in the annual cycle.

~

I touched the sharp needles of a young jack pine and inhaled the fragrance of the volatile oils stored within the green strands. To an ecologist or an aesthete, it was a sensory but most simplified world: sand below, dense groves of even-aged pines, abundant blue sky. To the biologist, the few colors painted a landscape that signified how a narrow breeding habitat serves as a potential limiting factor in a key stage in the annual cycle.

The host of the warbler, the jack pine (*Pinus banksiana*), is a native of the Far North, a boreal species adapted to the fierce winter wind and cold at the edge of the tundra. Its native range in Canada runs east of the Rocky Mountains from the Northwest Territories to Nova Scotia. The jack pine creeps into the United States from Minnesota to Maine. The southernmost extent of the range is northwestern Indiana, but distribution of the pine is spotty that far south. It begins to be more common only in the forests surrounding the northern Great Lakes.

Kirtland's warblers are plucky, but they are no tundra dwellers. Thus, much of the jack pine range now lies in a belt too far north, where only a few warbler species breed. The southern limit of the jack pine range—the only place thought to be suitable for this diminutive songbird—represents life on the edge. Some biologists have taken Seuss's "limited niche" hypothesis a bit further. They suggest that the best niches are located in the center of a species' preferred habitat, whereas poor-quality niches line the outskirts because, it is hypothesized, these edge sites blend ecological features from adjacent habitats that offer less of the right kinds of food, climate, or shelter. But to draw conclusions about this potential contributing cause of rarity is conjecture in the case of the Kirtland's; the small numbers of breeding birds preclude testing or arguing that the present distribution is worse because it is on the limit of its historical distribution.

Still, something doesn't compute. The range of the jack pine is huge, yet the Kirtland's remaining stronghold is a geographic splinter—about 160 kilometers long by 100 kilometers wide, illustrating just how range restricted this finicky warbler is. Could it be that the Kirtland's, like so many other species, was more common and its range larger during the last ice age than it is today? Back in the Pleistocene epoch (2.6 million to 12,000 years ago), when jack pine forests were extensive across the northern United States and southern Canada, Kirtland's warblers may have been far more common than they are today. But habitats shift with climate. For example, about all of the Kirtland's present-day range was covered by solid ice as late as 14,000 years ago. At that time, the range of the jack pine and its dependent warbler was probably a contiguous swath between the Appalachian Mountains and the Great Plains, probably well south of the edge of the ice.

After the retreat of what is known as the Wisconsin ice sheet about 14,000 years ago, the jack pine began to move north. But the Kirtland's warbler could not follow the cold-hardy jack pine all the way to the Northwest Territories, reaching only as far north as northern Michigan. Was the Kirtland's warbler a softie, a cold-intolerant species compared with others in its family? Both Carol and Sarah doubt it. After all, the Kirtland's has a greater body mass than many of the smaller warblers that nest farther north; in fact, next to the yellow-breasted chat, it is the largest warbler in North America.

Carol and colleagues think they have found the answer to this mystery. In the late 1980s, they explored the Seney National Wildlife Refuge's vast acreage of jack pine forest on Michigan's Upper Peninsula. They found lovely tracts of young jack pines, but the ground cover and soil moisture were all wrong for Kirtland's homesteading. Thick mats of moss covered the soaking-wet ground. Some high knolls did have the requisite blueberry and other shrubby ground species, but only in isolated pockets. The researchers were not surprised to find the young jack pine stands void of Kirtland's warblers. More detailed surveys on the distribution of very sandy soils farther north could support this conjecture with data showing that although pines are available, the essential sandy, well-drained soils are themselves rare.

At this same time, Burton Barnes and his team of graduate students were studying potential breeding areas of the Kirtland's warbler around Grayling and a few sites on the Upper Peninsula, documenting the specific features that the warbler seemed to prefer. Well-drained soils, both glacial outwash and areas of former glacial cover, seemed to predict Kirtland's warbler occupancy. More important to the bird's reproductive success may be Grayling sand or any similar dry, sandy soil and good drainage under the ground nest. As it turns out, the jack pine is the only tree tough enough to survive centuries of such depleted, dry soil conditions.

This toughness is epitomized even in the jack pine's cones. The next day in the grove, I grabbed a few to examine more closely. The small spheres looked as if they had been glued shut. Nature by design: jack pine cones are serotinous, a term used to describe seed-holding packets of plants that open only when triggered by fire. The cones can remain sealed for decades until a forest fire whips through a stand. The hot fire kills the mature trees but guarantees rebirth: when the heat of the flames hits 50°C (122°F), the cones pop open, spill the pine nuts in their protective shells, and reseed the charred earth. Nature's baton cues a new melody, for which the Kirtland's warbler is so precisely tuned, a firebird suite.

Historically, naturally occurring wildfires swept through the region and shaped the distribution and sizes of jack pine forests. In the nineteenth century, loggers came here in search of the coveted old-growth white pine for its fine, knot-free building timber but ignored the lowly jack. Intensive timbering ended in the 1880s after loggers pillaged the last stands of old-growth white pine and eastern hemlock. Extensive burning continued, which likely increased the range of the fire-adapted jack pine. This interlude, after the loggers cleared out but before settlers came, may have been the golden age in modern times for the Kirtland's warbler. The largest extent of jack pine habitat in Michigan probably existed between 1885 and 1900. After that, extensive fires either no longer occurred or diminished as human settlements spread. As jack pine stands matured under fire-prevention schemes designed to protect public and private property, the warblers, according to local and national surveys, went into a steep decline. Smokey the Bear and other popular symbols of the time may have prevented forest fires, but the suppression of fire was surely a contributing factor to the Kirtland's nosedive.

~

Visiting the breeding grounds during the Kirtland's spring pilgrimage is permitted only in the mornings, so birders must seek afternoon diversions elsewhere. The nearby Au Sable (pronounced O-sible) River offers fine canoeing and top-rated fly-fishing. For searchers of rarity, there is another one-of-a-kind, must-see site that in its own way deserves equal billing with the Kirtland's: about thirteen kilometers northeast of Grayling lies Hartwick Pines State Park, home to the largest remaining stands of 350-year-old eastern white pine and eastern hemlock in North America. And what could be a better ecological complement to a 5-year-old jack pine grove than a forest of centuries-old conifers? My curiosity got the better of me because some biologists speculate that in rare habitats one should naturally find rare species. What treasures did Hartwick Pines hold?

The walkway to the park was full of singing male rose-breasted grosbeaks, perhaps rivaling the scarlet tanager and the painted bunting as the most beautiful songbird in North America. Feeders at the reserve headquarters pulled in evening grosbeaks by the dozen.

A true deep-forest birding adventure awaited those who, like me, had never set foot in an old-growth white pine forest.

Nature writers have compared the feel of wandering into an old-growth stand to that of entering a cathedral. In the park, towering trees of massive girth created a cool environment. The understory layer was empty, as the dense shade prevented much of a shrub layer. The eastern hemlock was once called the "redwood of the east" because it can grow more than 50 meters tall and have a trunk 2 meters in diameter. In Great Smoky Mountains National Park, some 500-year-old trees still survive, but hemlocks elsewhere are not found in the large numbers they once were. Back in the 1920s, a bug from Japan known as the hemlock woolly adelgid appeared in New England and the mid-Atlantic states and began feeding on hemlocks of all sizes. Within eight years after infestation begins, a tree can die. Some ecologists say that the diminution of this widespread and once common hemlock may have had greater secondary effects than the loss of the American chestnut in the 1930s. Hemlocks play a vital role as habitat for species ranging from beetles to trout by providing deep shade along streams and cool refuges in the forest. Remove the hemlocks and the bright sun drives away the shade seekers.

Along the route I heard other warblers singing: pine, black-throated blue, and black-throated green. The ethereal songs of the wood thrush and hermit thrush made me stop and lie down under a grove of ancient trees to take it all in and reflect on what I had learned so far. I had filled pages of notes, but the lode of natural history insights danced around answers to some central questions: Why do species evolve to become habitat specialists in the first place? How widespread is this condition of extreme habitat specialization in nature, and is it more commonly observed in plants than in animals?

The answer to all three questions invokes evolution by natural selection. Imagine a species with a broad habitat tolerance. Some of its members will be breeding in widely distributed habitat types, while others will be reproducing in rare habitat types. If those individuals that reproduce in the rare habitat are to evolve into a new species specifically adapted to that habitat, two conditions must be met. First, they must become isolated from the rest of the population so that they breed only with one another, as we saw in the Foja Mountains of New Guinea. Interbreeding with individuals reproducing in other habitats prevents the evolution of habitat specialization. The second condition is that the reproductively isolated population must persist long enough for evolutionary processes to yield a considerably different organism, that is, a new species.

For several reasons, these conditions are much more easily met with plants than with animals. First, among plants whose flowers are animal pollinated and whose seeds also are disseminated by animals, genes are generally transferred over only short distances. In contrast with most animal populations, a plant population does not need to be far from others of its species to achieve the grist of speciation—reproductive isolation. Second, whereas few vertebrates can reproduce asexually, many plants can self-pollinate. This enables them to reproduce successfully in small, isolated areas. Third, because plants are intimately tied to the soil for their nutrients, they more readily evolve adaptations to specific soil types. Few animals have intimate relationships with particular soil types, and for those that do, the relationship is with structure (ease of burrowing, chances of becoming waterlogged), not chemistry. Thus the world abounds with plants that are restricted to specific soil types, often ones found in only a few places. Southwestern Australia, which features rocky outcrops separated from one another by only short distances, for example, has thousands of plant species found on only one or a few outcrops. Even those short distances are too wide for pollen or seeds to be transferred between them.

As I rested, my gaze turned to a cluster of lovely pink lady's slipper orchids growing in the sandy soil. Among the 250,000–425,000 vascular plants on Earth, perhaps as many as 20–30 percent can be considered extreme habitat specialists. To find many of them, though, I was sprawled out in the wrong part of the world. Because the spot where I was reclining, together with much of the northern forest zone, had been under a glacier 14,000 years ago, the habitats of this region are now broadly similar and the majority of temperate northeastern US forest species are relatively widespread.

Globally, extreme habitat specialization among plants is more prevalent in Mediterranean-like climate regions than in, say, the temperate forest of Michigan. Aside from one such region mentioned earlier, southwestern Australia south of the city of Perth, there are four others in the world: the California-Mexico coastal region; the Chilean Matorral; the Cape region of South Africa (also called the fynbos); and, of course, the Mediterranean region itself. The Mediterranean region, which offers one of the world's most benevolent climates for humans, is full of plants with narrow ranges because habitat conditions there are often extremely varied. In all five regions, characterized by winter rains and hot, dry summers, fire is an important part of the ecosystem, and the interaction of fire, climate, and different soils, many of them poor, generates a hotbed of habitat extremism.

The soils of Madagascar, New Caledonia, and other ancient islands that formed part of Gondwanaland often contain concentrations of heavy metals. Soils contaminated with heavy metals create a challenge for plants—much more so than for birds or mammals—but also present an extraordinary opportunity. By overcoming nutritional and water constraints of unusual soil environments, an evolving plant genotype escapes the customary tangled mass of vegetation and colonizes what may be virtually free space, with the benefit of reduced competition. For example, in the area around Cape Town, South Africa, the naturally rare plants are not a random bunch biologically. They are for the most part species that are killed by fire but whose seeds germinate in the burned-over soil and, like our jack pine, have short dispersal distances. South African biologists believe that this biology predisposes lineages to produce rare species. George Schatz, an expert on the flora of Madagascar, estimates that a considerable number of that island's

10,000 endemic species are strongly associated with a specific substrate, just as the Kirtland's warbler is to its sandy soils.

A final category of extreme habitat specialization in plants is that made up of climate refugees. These are species with a long history that have become "trapped" in refugia, a term biologists use to indicate a location of an isolated or relict population of a once more widespread plant species that persists today only in pockets where the microclimate still resembles what the plants experienced when they evolved. Worldwide, this group is of huge importance and includes many species. The widely planted dawn redwood (genus *Metasequoia*) is today represented by only 6,000 individuals in nature but was once one of the most common and widespread trees in forests across much of the Northern Hemisphere. The same is true for cycads, a group exemplified by rarities that we know as houseplants such as *Zamia* and those in the genus *Cycas*. Cycads are ancient palmlike seed plants that were extremely widespread in the age of the dinosaurs. More than 20 percent of the nearly 300 cycads are ranked as critically endangered or endangered. Although cycads live a long time, they reproduce infrequently and now have such small populations that further habitat destruction or theft for the plant trade puts some species at great risk of extinction.

The key point, I realized, is that substrates have a huge effect on plants, which typically cannot disperse their seeds far from the parent soil, in contrast with the dispersal capabilities of many animals. Thus globally, either unique or particularly inhospitable substrates, combined with widely varying climatic conditions of rainfall and temperature, yield tens of thousands of range-restricted plants whose numbers and percentages far exceed those of range-restricted vertebrates.

Yet, as we have seen, animals also may evolve to be habitat specialists. Some habitats in which an animal population has found itself have been sufficiently isolated that gene flow was prevented, and the area was large enough for the population to persist. How much separation is needed to prevent gene flow and how large an area is large enough depends, as we might expect, on the animal. Snails and insects, for example, move only short distances and, being small, often have high population densities. The animals that need the greatest distances and largest habitat patches to become isolated and to persist in isolation long enough to form new species are birds and mammals. That is why there are so few extreme habitat specialists in these groups.

The roster of birds and mammals that share the Kirtland's proclivity for extreme habitat specialization—reliance on a single habitat for some crucial aspect of their life history—is a relatively short one, but it is filled with a lot of exceptional and many famous creatures. Among the warblers, one could list the golden-cheeked warbler of the short oak groves on the Edwards Plateau in Texas and the golden-winged warbler of the eastern broad-leaved forests. Beyond warblers, perhaps no more than 3 percent of the 600 species of birds that breed in North America could be considered extremists. Best known are the northern spotted owl and the southern population of the marbled murrelet, both dependent on old-growth forest, and in the southeastern states, the red-cockaded woodpecker of longleaf pine forests that have the right mix of mature stands and fire to burn out the understory.

Among the roughly 400 species of North American mammals, about the same proportion, 3 percent, could be called extreme habitat specialists. Among the best known are the manatee of warmwater bays, the walrus and polar bear of the ice floes, and the pronghorn antelope and Utah prairie dog of the short grasslands. Globally, the most famous extremist is the giant panda, a species so specialized in diet that it is limited to forests with a dense bamboo understory. Many grassland rodents are extremists but have such wide ranges and are so prolific that they are actually abundant, such as the naked mole rat of southern Ethiopia, Kenya, and Somaliland. Among the more than 5,000 species of mammals globally, extreme habitat specialization is widespread across taxonomic groups and found on every continent.

The focus on rarity may leave the impression that a narrow range, low abundance, or, in the case of the species mentioned here, extreme habitat specialization is equated with a judgment—of inadequacy or evolutionary failure. On the contrary, the Kirtland's warbler and many other extreme specialists are superbly tuned to persist in their preferred environment. The trouble is, those environments may change. Kirtland's warblers do well in their jack pine home but depend on the conditions of their narrow habitat to exist. As the work in the Seney National Wildlife Reserve in Michigan's Upper Peninsula suggests, even where young jacks occur, the soil and undergrowth may have to be just right, too. When conditions change, however, and the forest stands age, Kirtland's warblers are forced to pick up stakes and move to the next patch or perish.

The firebird needs fire, and lots of it, but in the right places and at timely intervals—or, in its absence, carefully managed harvesting in jack pine stands to mimic the effects of fire. Yet, as I soon learned, there is more to the story of the Kirtland's rarity. Even if careless campers or closet arsonists were to set the sandy areas of northern Michigan ablaze—leading to more habitat—the species would still face an even greater threat.

~

On my second day in the grove, a new birdsong filled the warming air. It wasn't the explosive melody of the Kirtland's but a more metallic and bubbly offering. The members of my birding party noticed that the songs grew louder until we found the source. Hidden behind a natural screen of three-meter-tall jack pines sat a large cage containing six pairs of brown-headed cowbirds. On top was the entrance, accessible to birds wishing to enter and join the cowbirds inside, who were a most sociable lot. These, however, were bait birds, kept alive to attract others of their kind and leave the Kirtland's warbler in peace. For the unsuspecting cowbirds that entered this cage, there was no exit.

More than 100 species of North American songbirds, but especially breeding pairs of warblers, thrushes, and vireos, suffer from nest parasitism by brown-headed cowbirds. Kirtland's warblers are an especially easy mark for them. The deed is quick, the effects long lasting. When a female Kirtland's begins to lay a clutch of brown-and-white-splotched eggs, she steps briefly off the nest to feed. By the time she returns, a female cowbird, likely having cased the nest from a lookout post, will have visited and deposited an egg or two of her own, typically removing a host egg in the process. The cowbird's chick hatches a day or more before its adopted nest mates and grows more quickly. The adults fail to discriminate against the monster chick in their midst. They keep feeding the imposter, whose larger size means it gets more parental attention than the warblers' own young. Sometimes the more aggressive cowbird chick will push the rival offspring out of the nest.

For the poor Kirtland's warblers, much depends on how many eggs the cowbird lays in their nest; if only one, the pair have a chance to raise some of their own. If there are two cowbirds, none of the Kirtland's chicks survive to fledging. Unfortunately, female cowbirds are prolific egg layers. Over the May to July breeding season, one individual could conceivably put one egg in each of forty nests. This behavior may sound like the handiwork of an exotic species, able to exploit naive native birds, but while the brown-headed cowbird is a relatively recent arrival to Michigan, it is no foreigner. It is a North American species native to the South and West, where it was once known as the "buffalo bird." When hunters killed off the bison and ranchers replaced them with cattle, the cowbirds followed the domestic livestock too, feasting on the insects churned up by the hooves of the large grazers drifting through the grass. When logging and agricultural development began in earnest in Michigan in the eighteenth and nineteenth centuries, cowbirds followed, later spurred perhaps by the sharp decline of the bison in the 1880s.

How has the Kirtland's warbler held up under this home invasion by cowbirds, whose chicks hatch first and outcompete the natural nestlings for parental attention? Why has this invasion greatly affected rare species such as the Kirtland's yet abundant species hardly at all? To the second question, the short answer seems to be that some bird species have evolved ways to resist the nest parasites. American robin parents, for example, detect the strange eggs and roll them out. Others may abandon their nest and lay more eggs elsewhere or build another nest on top of the cowbird egg. Kirtland's warblers lack such behaviors, perhaps because, compared with these other birds, they have not yet had the time to evolve better egg recognition. Or the small size of the Kirtland's population may have reduced genetic variation available for selection of the defensive behaviors that have evolved in species with much larger populations and gene pools. Or both. In any case, the impact on Kirtland's reproduction grew to alarming levels. At its nadir in the early 1970s, less than one-third of warbler nests in the Grayling population produced any young, largely attributable to the influence of cowbird parasitism.

In 1972, the US Fish and Wildlife Service, along with other federal and state agencies, began controlling cowbirds with large live traps, such as the one we'd seen, placed in Kirtland's warbler nesting areas during spring and early summer. Wildlife technicians check the cages daily and euthanize any trapped cowbirds—an average of 4,000 per year.

"No one wants to kill cowbirds, but the trapping has worked a small miracle," Sarah Rockwell related. Nest parasitism rates had dropped sharply, from 69 percent in the late 1960s, before trapping began, to less than 5 percent. Even more inspiring, average clutch size had increased from 2.3 eggs per nest to more than 4, and the average number of young Kirtland's warblers fledged per nest increased from fewer than 1 to almost 3 birds during the same period. Cowbird rustling is an expensive task, and euthanization seems gruesome to some. But for the attending biologists, watching the last few Kirtland's warblers unwittingly feed supersized cowbird chicks seemed at first surreal and then shameful. Cowbirds have their right to live, too, but at what point do we intervene when their hardwired behavior threatens to drive other species to extinction?

Over coffee with Sarah, I posed a hypothetical question about the Kirtland's future: "If half of northern Michigan were turned into a jack pine preserve and immature stands were increased by a factor of ten, through either fire or management, would we still need cowbird control?"

"Well," Sarah replied, "what you need to answer your question are data from the last time we had reproductive success information collected before cowbird control." Larry Walkinshaw, the dean of Kirtland's warbler biologists, had earlier estimated reproductive success to be 0.8 fledglings per pair of adults per year. "So, in the absence of cowbird removal, I guess I wouldn't expect success to go to zero, but the area could certainly become a population sink." The term Sarah used describes a state in which deaths exceed "recruitment," or addition of young to the population. And then there is the sad truth: cowbirds are nomadic, so removal during one year has no bearing on the number of cowbirds that will arrive the following year. Nest parasitism would rise again within a year or two and warbler numbers would dwindle if biologists stopped the trapping program.

~

The first federal Kirtland's warbler recovery team, established in 1973, realized that for a bird with such a limited range, it was essential to create more areas of suitable jack pine for it to breed in. So, during the mid-1970s, the recovery team designated some 540 square kilometers of jack pine stands, spanning state and national forests, for management as Kirtland's warbler nesting habitat. Additional lands were added during the 1990s and in 2002 to bring the total public land area specifically set aside for the Kirtland's warbler to more than 770 square kilometers. But how could these and even larger tracts of jack pines necessary to underpin a recovery be maintained without burning huge areas, even some areas close to towns?

Silviculturalists from the USDA Forest Service argued that fire wasn't the only management tool in the box. Logging practices and plantations could create conditions acceptable to breeding warblers, they said. Ecologists were skeptical. In their view, forestry experts always suggested logging and intensive management as the solution to any conservation problem. The critical question was whether Kirtland's warblers would breed and survive as effectively on plantations as in naturally burned forests.

Again, Carol Bocetti's dissertation work provided answers. "I found that plantations had lower jack pine density and fewer openings in the managed forests than in natural wildfire areas," she said. But, remarkably, she discovered that the critical response by the warblers—measured by male density, clutch size, number of young fledged, and number of parasitized nests—did not differ between plantations and fire-maintained habitat. It was all about where the females were. Still, she recommended increasing both jack pine density and the number of openings in future plantations, and managers have done so since then.

For a dedicated conservation biologist, a published paper is one reward. But the best outcome is seeing one's research findings become management policy and then witnessing that policy's positive effect in helping to save a species. "I think Carol's work shows that the logging and replanting are effective in replicating fire-maintained habitat," Sarah concluded. At least for the Kirtland's warbler, Carol resolved the controversy through research. Almost all Kirtland's habitat is harvested and replanted now, so there is very little variation in regeneration type to study. Little is yet known, however, about other vertebrate and plant species living in the jack pine forests that do require fire episodes.

A footnote to the Kirtland's story made me hopeful about the species' long-term survival prospects. Until 1996, all known nests had occurred within sixty miles of Grayling. But in 2007, three Kirtland's warbler nests were discovered in central Wisconsin and one nest was found at a Canadian Forces base in Petawawa, Ontario. By 2010, the presence of 25 Kirtland's males in four Wisconsin counties had been recorded, and fifteen nesting attempts were monitored that summer. And one of our guides remarked that the number of singing males in Michigan's Upper Peninsula had increased to 34 in 2010. Sarah thinks these developments are a sign that the population of nearly 2,000 singing males has grown to near saturation of their habitat in the Grayling area and has started expanding into new areas.

An ability to disperse from a breeding or feeding area that has somehow changed in quality or become overpopulated to an equal or superior but more distant new site makes you what biologists call a successful fugitive species: when conditions shift over time, you escape and find new ground.

Unfortunately, most other extreme habitat specialists, especially plants on unusual soil types and many terrestrial invertebrates, are relatively poor dispersers or have nowhere to go. Unlike the Kirtland's warbler, they are incapable of dispersing when their habitat begins to change. Often they have resided in habitats that have been stable for a long period of time, long enough for them to be seen as distinct endemic species and perhaps intolerant of different habitats nearby. Many of the plants in the Cape flora of South Africa, in southwestern Australia, and in New Caledonia are in this situation. There is an old German expression, "Never move an old tree," a metaphor for the unintended consequences of shifting the aged from their longtime domiciles to strange locales. The literal meaning of this phrase also holds true: most trees, shrubs, and herbaceous plants can't move far from their home base, nor can the flightless or weak-flying insects that live on them. In this way, lack of dispersal ability both contributes to isolation and sets the stage for the emergence of new species, many of them rare.

On the drive back to Detroit, not far from Grayling I pulled over by a large marsh where, in late morning, the songbirds were still active.

The imperative of establishing and holding territories in mid-May leaves the abundant yellow warblers and yellowthroats with little time for rest. "*Sweet-so-sweet, I'm so sweet*" and all its variations for the yellow warbler, "*Witchity-witchity-witch*" for the common yellowthroat: an orchestral piece for two warblers, with encore after encore. There were no Kirtland's warblers here; they shun such swampy locales. But it was a happy scene, a wetland brimming with song and color.

The yellow warbler and the common yellowthroat are unlikely ever to have a national task force and recovery team formed just for them. In many ways, these two warbler species are the opposite of the Kirtland's. Both the yellow and the yellowthroat have massive ranges, among the largest of any North American songbird, and unlike the Kirtland's they venture far north, into Alaska and the Yukon, respectively. In this Michigan marsh, they were singing in the center of their range—the safest place to be, as the rarity theorists tell us. On top of its range security, the yellow warbler has evolved rather rapidly to outmaneuver the cowbirds. Yellow warblers abandon nests or just build a new nest lining over a parasitized clutch and lay more eggs. These birds are poster species for resilience and resistance to extinction. By contrast, in Michigan Kirtland's warblers will likely remain wards of the state. They represent what biologists call a conservation-dependent species.

Can the recovery efforts for the Kirtland's warbler be applied to other species that are extreme habitat specialists? For the Kirtland's, ensuring the existence of sufficiently large landscapes containing the right mix of young jack pines with other forest classes and controlling the numbers of parasitic cowbirds are key. Because of the cowbird infestation and its own extreme breeding habitat requirements, the Kirtland's joins the ranks of other conservation-dependent extremists such as the rhino in Nepal. For this species, habitat protection, restoration, and antipoaching measures are essential. The recovery strategy for more mobile extremists may offer less insight for the sedentary rarities, especially plants and invertebrates. For these extreme habitat specialists, we have to map their ranges and save them where they occur.

From John Terborgh's studies at Lago Guri described in chapter 3, we saw the impact of the loss of even low-density jaguars, the top predator, in areas of Venezuela: the herbivore populations went haywire and defoliated many plants. Other keystone species, too, can play important roles as ecosystem engineers. Of what ecological effect is a bird so small and so few in number as the Kirtland's, or other rarities like it? Kirtland's warblers, or any of the hundreds of other migratory songbirds—warblers, thrushes, vireos, orioles—cannot be viewed in isolation from their native ecosystems, the broadleaf and conifer forests of North America. We know that Kirtland's warblers gobble defoliating insects and their larvae on their breeding grounds. But even in their heyday, this species was never abundant. However, if we total the original number of all migratory songbird species, many of which are now in decline, their ecological impact as consumers of leaf-eating caterpillars must have been enormous. The ornithologist Scott Robinson has noted that for warblers and other songbirds to play their traditional collective ecological role in forest communities of keeping insect outbreaks in check, they need to be abundant. Today, Bachman's warbler is extinct, Kirtland's is down in the danger zone, and birds such as the cerulean warbler are nearly as threatened. Will there be few or no checks and balances in a brave new ecological world?

The shape of that brave new ecological world will be strongly influenced by climate change, including temperature increases, sea level rise, and shifting patterns of storms, precipitation, and drought. A century or two from now, will that narrow nesting habitat preferred by the Kirtland's warbler even be there? If the range of the jack pine extends much farther north, why can't the Kirtland's just shift up a few degrees latitude? Remember that jack pines can grow on different soils, but the sandy soils preferred by this extremist may not map to the jack pine. In the case of the Kirtland's and other species that migrate to the Bahamas, several climate models predict drier conditions in the archipelago that could reduce the supply of fruit in late winter, while predicted sea level rise could put some wintering areas under water. How these new conditions will affect the range and density of this species and other rarities favoring fire-dependent habitats is unclear.

For the Kirtland's warbler, persisting on the edge of its former range seems like a ticket to becoming a climate refugee. Yet perhaps more catastrophic forest fires—another prediction of some climate change models—might create more breeding habitat for this bird. Then again, global warming might cause the jack pine's range to shrink from the south, because jack pines thrive in cooler areas, and the species so far has retreated north since the last glacial maximum. Now that we know the world's climate is changing even faster than during the last ice age, it also becomes apparent that species that are extreme habitat specialists, such as the Kirtland's in the jack pines, may not be able to adapt fast enough to deal with changing ecological circumstances.

The story of the Kirtland's warbler and other habitat specialists like it offers an important insight into the nature of ecological rarity. The conditions we observe today are unlikely to be similar either to those that existed when our current habitat specialists evolved what seem to us to be strange peculiarities or to those that will obtain fifty or a thousand years hence. Thus, the Kirtland's warbler may have evolved its adaptations to young jack pine stands when that habitat was very common. Today that habitat has greatly shrunk in extent, owing to both historical climate change and human activities, such that the bird now appears to have extreme requirements. In other words, many of today's habitat specialists may have been common and widespread in the not-too-distant past. Conversely, many of today's common species may become rare habitat specialists in the near future. For example, before Europeans colonized the area, clear-cutting forests and starting many fires, old-growth conifer forests dominated the mountains of the Pacific Coast of North America. The favored habitat of the northern spotted owl was, in fact, the dominant vegetation type. Species that preferred scrublands and young forests, such as the chestnut-sided warbler, in the same genus as the Kirtland's, were likely rare. Had we visited the area then, we would have called them, not spotted owls, habitat specialists. A striking contemporary example of this kind of change is provided by a recent National Audubon Society survey that shows that populations of many common North American birds are rapidly declining. We do not shoot or poison most of them. Rather, we are driving them toward greater rarity by making their preferred habitats increasingly rare. And if we lose polar bears, it will be because climate change made their required hunting habitat—sea ice—too rare.

Many species are in fact, to one degree or another, habitat specialists. Think, for example, about the several species of prairie dogs that once were among the most abundant and widespread vertebrate species in the American West. Prairie dogs can only live in short grasslands, so they are true habitat specialists, but their habitat was once so abundant that they occurred at high densities over a vast range. For most species, the condition that enable them to survive and reproduce at high enough rates to persist over time are quite specific. As habitats change, for whatever reason, species may have difficulty adapting; some may go into steep decline, while a few may thrive. Thus, we may be misled if we think of today's habitat specialists as being different from species we call habitat generalists. All species have some specific habitat requirements, and we need to be alert to changes that may substantially affect their fate. Currently rare species such as the Kirtland's warbler may require special and urgent attention, but in the long run, all species are vulnerable.

Chapter 5

There in the
Elephant Grass

O N AN AUSPICIOUS DAY IN APRIL 1974, circled by court astrologers, Nepal's royal family gathered for a coronation ceremony in the capital, Kathmandu. In a palace chamber, Brahmin priests wrapped Crown Prince Birendra in gauze and then enveloped him in the skin of a male greater one-horned rhinoceros. The disguised monarch-to-be entered the throne room to chanting and the burning of incense. In choreographed motions, he began shaking the rhino's meter-long penis at his kin; then he repeated the dance ritual. Witnesses were few, but the performance evidently won celestial approval. The gods were praised, the royal family duly felicitated. Birendra stepped out of the skin suit; the kingdom was his for the taking.

Among traditional cultures, many rare species are believed to bestow special powers on those who eat them or wear their body parts. The rhinoceroses fall into this category; they have intrigued humans East and West for centuries and have been the subject of mythology, awe, and terrible persecution for their horn. For the West, rhino expert and wildlife historian Kees Rookmaaker explained the origins of this allure. "The Western world became acquainted with the rhinoceros in 1515 from Albrecht Dürer's anatomically inaccurate woodcut. It was drafted from notes and a sketch of a captive greater one-horned rhinoceros brought to Portugal from India." Because the subject was large, dangerous, and rare, the woodcut attracted wide attention; it also reawakened the legend of the unicorn.

Known to science as *Rhinoceros unicornis*, the greater one-horned rhino bears little resemblance to the magical horned horse of fairy tales. This is a mammalian titan, more tank than prancer, a massive beast covered in what resembles armor held together with welds (skin folds) and rivets (tubercles). The greater one-horned rhino stands two meters tall at the shoulder and is the fifth-largest land mammal on Earth. Its horn is not technically a horn but rather densely compacted hair fibers pressed into a pointed cone. Whatever it is called, the protrusion has been supposed to have magical properties: it has been prized as medicine to quell life-threatening fevers, rumored in Vietnam to cure cancer, and nearly everywhere erroneously believed to elevate libido. In 2012, rhino horn was worth more than its weight in gold, and its value skyrocketed to $100,000 a kilogram. For this object, rhinos have been a favorite target of poachers, even though, as one wag put it, ingesting rhino horn has the same medicinal value as gnawing one's fingernails, also made of keratin.

None of the claims hold up to scientific scrutiny, but myths linger. A most stubborn belief surrounds the value of rhino urine, known as *muth*. Drinking the foul liquid, which fetches five dollars a liter on the open market, supposedly cures asthma and tuberculosis, while applying a few drops is said to heal inner ear infections.

My elephant driver, Gyan Bahadur, and his colleagues routinely jumped off their mounts to gather up the spilled urine in a plastic bag when a rhino happened to urinate in front of them.

We've now moved from the cool jack pine forests of the northern United States to the steamy lowland jungles at the base of the Himalayas to investigate several new causes of rarity. The 14-gram Kirtland's warbler was rare because of limited breeding habitat and nesting disruption by cowbirds. But this bird can potentially nest anywhere across the jack pine belt of the northern continental United States where the trees are young and the soil is sandy. The 2,000-kilogram one-horned rhinoceros, in contrast, is much more range limited, rarely wandering farther than two kilometers from water and feeding intensively on the thin strip of wild sugarcane that borders the major rivers of lowland Nepal and northern India. The Kirtland's, like many other rarities, evolved as a fire-dependent species. The greater one-horned rhino and some other narrow-range species evolved as floodplain specialists—those that persist only close to the river's edge.

Another difference between the two habitat specialists is the level of competition with humans: agriculturalists covet the silt-rich grasslands found along the riverbanks and have transformed them into the rice bowl of South Asia. In contrast, the nutrient-poor, rapidly percolating soils in the jack pine zone deter farmers from the thought of grain production. The two species lie at opposite ends of the demographic spectrum as well. A female Kirtland's typically dies by four years of age or younger, whereas a female rhino can live as long as fifty years.

The greatest differences and most important new causes of rarity to explore are human predation and breeding biology. As rare as Kirtland's warblers are, no Native American tribe ever coveted their feathers for headdresses as we saw New Guinea highlanders use bird of paradise feathers or Amerindians in Peru hunt macaws for the same purpose. Michigan homesteaders never used Kirtland's warbler meat, beaks, or claws as cures or aphrodisiacs. Poaching was simply never an issue, in contrast with what all rhinos face from Africa to Asia. And as we shall see, the critical demographic feature that dictates recovery or doom for a rhino population depends upon females living to a ripe age and producing as many calves as possible over their long reproductive lifetimes. This overlooked population parameter—known as adult female mortality—it turns out, holds the key to understanding and restoring populations of nearly all the world's charismatic large mammals, from whales to grizzly bears to giant pandas to elephants, not just the one-horned rhino.

As we saw in the Amazon, the rarity of top predators is governed by the basic laws of thermodynamics—there simply can't be many individuals in an area that live solely on the flesh of the larger mammals. The same principle or law should not, however, affect a plant eater such as the one-horned rhino, which has a massive fermentation vat attached to its stomach to digest the superabundant elephant grasses in its range. In fact, a distant relative of the one-horned rhino was the largest land mammal that ever lived. Neither mastodon nor prehistoric elephant, the heavyweight prizewinner is the extinct giant giraffe rhinoceros, as tall as a double-decker bus and six meters long.

As a young wildlife biologist studying the one-horned rhinoceros during the reign of Birendra, the rhino king, my first goal was to explore how large body size affected the ecology and conservation of giant mammals such as the greater one-horned rhino and how it influenced their range and abundance, our conditions of rarity. A second goal, as noted in chapter 1, was to determine whether big plant-eating animals are passive participants or ecosystem engineers of consequence even when they occur in low numbers.

In Nepal in the mid-1980s, one-horned rhinos occurred only in Chitwan National Park, 145 kilometers southwest of Kathmandu, in the lowland jungles known collectively as the Terai zone. After two years of trying, my colleagues and I finally received permission then from the king of Nepal to study wild rhinos. Leading the intensive effort was a group of park staff members, Smithsonian Institution

Map of Nepal and neighboring regions

biologists, trackers, and elephant drivers. The plan to save these rhinos included catching them and attaching radio transmitters so we could monitor their movements. The project was spearheaded by my collaborator, Hemanta Mishra, at the time Nepal's leading wildlife biologist and a close adviser to the king.

The insights into large-mammal biology in general and rhinoceros ecology in particular keep coming. Today, more than twenty-five years later, we know a lot more about this extraordinary animal and what it has to offer us in understanding the causes and consequences of rarity. But it all began with trying to catch the first one.

～

On a fog-laden morning in November 1986, Vishnu Bahadur Lama, our chief tracker, was out on elephant-back searching for a male rhino. Instead, he came upon a mother rhinoceros and her calf drifting along the riverbank. Vishnu motioned for the driver to push their elephant onward to find the male. On a high point along the riverbank, Vishnu stood up on the elephant's back for a better view. Game trackers out at sunrise heed the words of old-timers: *read the grass.* He scanned for a hole in the wild cane, for hidden in the gaps might be a sleeping male. Another tip: *watch for mynas.* The gregarious tickbirds move among sleeping rhinos to feast on skin pests. Vishnu strained to see a few black smudges rising up from the tall grass.

The rest of us hunched over a glowing fire, waiting for Vishnu's return. Our game trackers chatted as they took their continental breakfast of tea and biscuits. A short, bowlegged man poured steaming cups of Darjeeling thickened with sugar and water-buffalo milk. There was no rush; by nine o'clock, the lid of fog covering the Chitwan Valley would have burned off and the sun would peek through the silk cotton trees. Then the hunt would begin. Gyan Bahadur, an elephant driver for more than thirty years, was the group's resident sage, or at least he thought so. Gyan and his crew would soon be urging their powerful charges through the grass toward our quarry. Other dangers lurked in this jungle, though. One could surprise a tiger, a sloth bear, or even a king cobra, whose venomous bite could topple an elephant. Nearby, young men hustled to feed the research vehicles. They hauled piles of wild sugarcane and tree branches before a row of hungry elephants, which delicately selected choice bits of vegetation with their trunks and began a noisy repast: chewing, lip smacking, more chewing, and prolonged bouts of flatulence.

The banter halted when Vishnu and his mount hustled into camp. "A big male. With a horn like this big!" Vishnu spread his hands even wider than his grin. "He is the one." In seemingly one movement, he hopped down from the elephant, dropped to his haunches, and spread his palms over the fire. Gyan handed him a steaming mug. Some of the trackers were eager to start while the air was cool; an early capture would reduce the risk of an animal overheating in the noonday sun.

Finally, everyone was ready. Vishnu took a shortcut to the saddle: he grabbed the tips of his elephant's ears and held on while the beast offered Vishnu her trunk to use as an escalator and in one smooth motion lifted him to the top of her head. He scrambled around the driver and found a space on the broad saddle next to Sunder Shrestha, the veterinarian in charge.

Vishnu had begun his career here as a teenager, hauling water and chopping firewood. After a few years, Hemanta had made him a *shikari*, or tracker. Over time, he and other young men from his mountain village had become Nepal's best wildlife technicians. Catching, radio-collaring, and monitoring elephants, leopards, sloth bears, forest deer, and tigers—man-eaters or more orthodox deer eaters—were all in a day's work for these enthusiastic hill tribesmen.

The elephants marched single file to the edge of the broad Rapti River. The pachyderms squeaked and rumbled as they leaned into the current. Elephants speak in a subsonic language largely undetectable to the human ear. What we can hear has a deep bubbling quality, something like gastric distress. But it has great use as a means of communicating with other elephants over long distances, similar to how whale song transmits across the ocean. The white rhino and Sumatran rhino are now also known to engage in subsonic dialogue with other individuals of their kind. For large-mammal species that live at low numbers in dense habitat, such a trait must be of great use in locating one another, especially when females are receptive to breeding.

After crossing the Rapti, we fanned out in the tall grass. The elephants swept the area where Vishnu had found the sleeping rhino. A spotter saw it and waved his arm for the drivers to form a ring around the animal to prevent him from escaping once he was darted. A drugged rhino that staggered into the river would drown, and the fallout could scuttle our program before it even started.

Mel Kali moved in closer, carrying Vishnu and Sunder on her broad back. Faced with a charging rhino or tiger, a well-trained elephant like Mel will stand her ground, trumpet, and scare it off. The driver guided her to the sleeping rhino and stopped at a distance of twenty meters. Sunder took aim at the rhino's enormous rump and—*Pop!*—made a perfect shot. The red chenille tail of the syringe hung from the rhino's posterior. Snorts of indignation from the front end heralded an imminent charge, but the attack never materialized because the narcotic quickly entered the rhino's nervous system. Sunder's textbook on wild animal capture indicated an eight-minute lapse between injection and sedation. He checked his watch; everything was running flawlessly.

The rhino stood motionless as the drug worked its magic. I squeezed the radio collar I was holding and glanced at my watch, the seconds ticking away, until the rhino sank to the ground at the edge of the forest. Finally, we dismounted. "So far, so good," I thought. Before giving the green light to begin work, however, Sunder waved me and the others back, saying, "Let me test first for full sedation." Grabbing a stick, he poked at the rhino's broad backside. So much for the textbook: the two-meter-tall rhino jumped to his feet.

The elephants trumpeted in alarm. The men on the ground scrambled for safety, trying not to be trampled by the fleeing rhino or panicked elephants. Fortunately, no one was hurt and the moment of bedlam passed. Behind them, Vishnu, the other trackers, and the elephant drivers howled with laughter. The excitable chief park warden, Ram Pritt Yadav, joined in. Hilarity in the aftermath of near disaster was practically a national custom.

Sunder was already preparing another drug cocktail for a second attempt. The trackers gathered round. He explained, "Either the contents of the first syringe didn't inject at all, or, most probably, it only partially injected." This seemed like a plausible hypothesis, preferable to the alternative—that modern sedatives had no effect on ancient pachyderms.

The rhino was still in the vicinity, and within minutes we had repeated the drill: encircle, dart, wait until the drug put the rhino under its spell. This time it worked, and soon we were swarming over the sleeping male, covering his eyes and plugging his ears to minimize arousal and attaching the radio collar. "Forty-five centimeters," I said to Ram Kumar Aryal, who was writing measurements on a data sheet. I had just measured the massive horn, as long as a chair leg. We tested the collar's transmitter, and it sent out a strong signal. The beacon would allow us to locate the male with minimal effort even when he was in hiding and, after a trial period, enable us to habituate him to our presence on elephant-back. We could then learn where he lived, the pattern of his movements, and more intimate details associated with breeding.

Our team posed for a group picture, gathering around Yadav, as we named the rhino in honor of the presiding chief warden. Then it was time to saddle up and, from the safety of elephant-back, watch the rhino wake up. Vishnu stood next to the massive head and held the rhino's ear so Sunder could find a vein. The skilled vet administered the antidote and then both men remounted quickly. The drivers and trackers sat in awe of modern pharmaceuticals: a few drops of a drug a thousand times more concentrated than morphine to knock down a massive rhinoceros, a few more drops of the fast-acting reversal agent to right the beast. In less than thirty seconds, Yadav was rejuvenated. The big male shoved off, uttering deep, huffing grunts in cadence with its departing trot.

As we headed back to camp, one of the drivers burst into song and was joined by a chorus of colleagues. A celebration lay ahead. There would be feasting, folk songs and native dances, and a retelling, for at least the tenth time, of the day's big adventure. For me, the capture was burned into memory: on that foggy morning, there in the elephant grass, I had touched my first unicorn.

～

I continued to study this remarkable species for several more years in Nepal, and my fascination with rhinos continues to this day, even as they barely hold their own or, in some cases, hurtle toward extinction. In the 1700s millions of rhinos were alive, but by 2011 the numbers of the five living species together totaled slightly more than 28,000 wild individuals. Most common are the 20,140 southern white rhinos, followed by the 4,840 black rhinos, all in Africa. In Asia, there are 2,900 endangered greater one-horned rhinos. The Sumatran rhino, of which there are about 250 left, and the Javan rhino, with a population of fewer than 50, are among the rarest mammals on Earth.

A species can have a narrow range and low density and yet, as we saw with birds of paradise, bowerbirds, and even Kirtland's warblers, can still be easy to spot if you look in the right places. The same is true for three of the rhinos, the white, black, and greater one-horned. No such "good viewing habitat" exists for the other two species. Would that statement have been true two thousand or even fifty years ago? To a conservationist, the question might seem academic. They are rare now, so it is better to address the immediate, pressing issues of poaching and habitat loss. Yet, to craft a longer-term strategy, biologists need to know under what conditions rhinos might once have flourished, in order to understand how it might be possible for them to flourish again.

A little paleontology greatly aids our understanding. The fossil record shows that rhinos first appeared in the late Eocene epoch (35 million years ago), the earliest known rhinoceros-like mammals having been described from fossil deposits in Asia, North America, and Europe. They looked nothing like the giants of today. Instead, these primitive rhinoceroses resembled early horses and tapirs, rhinos' nearest living relatives. They were diminutive and delicate, were probably abundant, and lacked horns. Between the late Eocene and Oligocene (about 33 to 23 million years ago) perhaps the most dramatic change was an increase in body size from the ancestral forms to larger, more recent body types, a phenomenon frequently observed in mammals. This is an example of Cope's rule—that species within a lineage tend toward gigantism over evolutionary time. Another, more recent development was the appearance on the rhino skull of a unique, boneless horn. Two traits helped the rhinos to spread widely across the landscape. The evolution of broad feet with three toes allowed them to adapt to marshy habitats and expand their range. Of greater ecological importance were increases in the size and height (from the gumline) of high-crowned molar teeth to better handle a diet of coarse grasses.

Growing so large was probably a defense against natural predators. After rhinos reached a certain size, there were no longer any natural enemies large enough to kill any except the very young calves. Rhinos filled every herbivore niche possible except the arboreal, although the previously mentioned giant giraffe rhinoceros browsed the limbs of trees. The animal's feeding strategy was to eat a great deal but process the bulk quickly in its giant digestive sac. And being giant herbivores meant they were capable of making a living by eating lush, coarse elephant grasses. No matter how one defines evolutionary success—diversity of species, persistence over time, range size, biomass, or feeding niches occupied—rhinos dominated the early epochs when large mammals flourished, in the late Eocene and the Oligocene. On the basis of their numerical dominance, their biomass within an ecosystem, and the number of diverse, wide-ranging, but now extinct species that once flourished, we can safely say that many ancient rhinos, when compared with other mammalian browsers and grazers, were ecological commoners: widespread and abundant.

So what caused the eventual decline of this group? While large body size does not cause rarity, there is an important implication to it that can be drawn from studies of the mass extinction of large mammals during the Pleistocene epoch. The once abundant large plant-eating mammals—the mammoths, mastodons, and their kind—were especially vulnerable to environmental changes and relatively easy to hunt. Early hunting hit large animals especially hard. A major reason is that large animals, especially herbivores, inevitably have very low reproductive rates. Their persistence depends on high adult survival rates. Thus, even a slight increase in death rate can bring their numbers down rapidly. When humans, whether early hunters or latter-day poachers, take females from the population time after time, crash is not far behind.

In North America, almost the entire rhinoceros fauna was eliminated between 2.5 and 5 million years ago, and in Eurasia only two lineages survived. One of these gave rise to a form quite similar to the Sumatran rhinoceros of today. Perhaps the most famous member of this line was the woolly rhinoceros (*Coelodonta antiquitatis*), which appeared first in China and moved westward into Europe. This charismatic thick-coated, big-horned species inspired the early painters who decorated the walls of the Lascaux caves in France. It persisted until about 1,200 years ago and ranged from Korea to Spain.

A little ecological history clarifies the swift and widespread decline among all five living rhino species in more recent times. Before approximately AD 1400, when the Gangetic Plain first opened to agriculture, greater one-horned rhinos must have been relatively common in this hot, steamy region. Basing our calculations on densities achieved by rhinos today in prime floodplain habitat, I estimated that perhaps half a million or more grazed along rivers or wallowed in oxbows. Around 1900, there may have been 300,000 to 1 million black rhinos in sub-Saharan Africa. Several hundred years ago, the numbers of white rhinos must have been equally staggering. And Javan rhinos were so common in colonial Indonesia that they were considered garden pests and shot by Dutch tea planters. Even Sumatran rhinos then must have filled Southeast Asian forests with their peculiar whalelike vocalizations. All evidence seems to point to the conclusion that rarity is a relatively recent phenomenon for rhinoceroses and that overhunting, then poaching, and habitat loss have been the drivers of their startling decline.

～

In that September of 1986, before we had our first successful radio-collaring, torrents of rain had engulfed the Chitwan Valley for three solid days. Along the banks of the Rapti River, villagers gazed in wonder and fear at the surging brown floodwaters that swept uprooted trees along as if they were toothpicks. As I stood watching above the river's roar, the ecological significance of this event became clear. Floods occur here almost every year, but rare, catastrophic floods are the major structuring force in the one-horned rhino's home ecosystem, redrawing the vegetation map. Had I not been here to witness this rare event, I might have missed seeing how a once-in-a-century or half-century deluge inundates vast areas, reshaping the ecology of the land. Such a flood reconfigures floodplains and oxbows and even shifts the vegetation types over huge areas through the deposition of rich silt carried from the mountains.

Every year, monsoon flooding forces the wildlife of the Terai zone to head to higher ground or risk being swept away. When the threat passes, however, new life emerges. As each year's floodwaters recede, the silt deposited in the grasslands acts as a layer of fertilizer. From that silt, the next year's crop of wild sugarcane will emerge, and rhinos will feed on the shoots. The ability of greater one-horned rhinos to feed on early postmonsoon vegetation maintained by annual disturbance, to track those disturbances, and to disperse among them is vital to their survival.

Fortunately, river-hugging rhinos are powerful swimmers. They can easily ford the flooding waters and are savvy enough to avoid exposed quicksand when waters retreat. Not all of Chitwan's grassland-dependent mammals handle the floods with such aplomb, however. Endangered hog deer are sometimes washed away by the rising waters. The elusive pygmy hog and hispid hare, also called the bristly rabbit, are some of the rarest mammals on Earth and face emboldened predators when floods bury their tall grass cover. Yet such native species must have made some adaptation to the normal range of annual flooding and the once-in-a-century deluges, as species truly unable to cope would have vanished ages ago.

～

Today, one still finds very large mammals in the tropical belt where a deadly disease such as Ebola, malaria, or dengue fever is rampant, where the soil is infertile, or where both conditions leave the landscape better suited to wildlife than to human habitation and agriculture. In other places, such as sub-Saharan Africa, outbreaks of diseases such as rinderpest and sleeping sickness, carried by the tsetse fly, spread among livestock and their herders, preventing both from occupying natural habitats. Until 1950, a deadly strain of malaria kept the Terai zone habitat of the one-horned rhino in Nepal and India virtually free of humans and relatively intact. Before 1950, there is every reason to believe, rhino numbers in Chitwan exceeded 1,000 individuals, a healthy population by any count.

In the mid-twentieth century, several political and social changes in Nepal and India indirectly contributed to the rhinos' steep decline and then to their beginning recovery. First came the fall in the early 1950s of the autocratic Rana kings who had ruled the country for one hundred years. The Ranas' overthrow reversed the policy of using the Terai forests as a malaria-ridden barrier to ward off would-be invaders (the British) from the south. The threat of invasion disappeared with India's attainment of independence from Britain, and beginning in the late 1950s, malaria eradication programs in the Terai—financed largely by foreign aid agencies—brought the scourge under control. This opened the Chitwan Valley to settlement by the tens of thousands of impoverished hill farmers who streamed in from all over Nepal. Unfortunately for the greater one-horned rhino, its preferred feeding areas, covered in wild sugarcane, also made the most productive rice paddies. The human population in this region rose from 36,000 to 100,000 in a decade. By 1960, the entire length of the 160-kilometer-long valley was inhabited, and most of the forest and grassland habitat had been converted to a brilliantly colored mosaic of mustard, rice, and maize. Similarly, malaria control in India across the entire Terai belt, including along the Ganges and Brahmaputra Rivers, spurred rapid land conversion. In a few decades, the Terai elephant-grass ecosystem shrank to less than 2 percent of its original range.

Other threats emerged to the rhinos' well-being, as well as to that of Chitwan's other famous charismatic large mammal, the tiger. By the late 1950s, hunters had dispersed throughout the Chitwan Valley, poaching the rhino for its valuable horn and tigers for their bones and skin and the male's penis. Before the eradication of malaria in the area, the density of tigers must have been staggering, perhaps among the highest in the world. During a royal hunt led by the Rana kings of Nepal in the winter of 1938, during the two-month window when the malaria-carrying mosquitoes were inactive, 125 tigers were shot in a relatively small part of Chitwan. To put this in perspective, the number of tigers shot in that narrow section of no more than 50 kilometers long and a few kilometers wide in 1938 equals about half the number of adult tigers occupying, in 2012, what is now the Terai Arc Landscape—a 1,000-kilometer stretch of eleven tiger reserves and connecting jungles across lowland Nepal and adjacent India.

With continued settlement and expanded rice cultivation, the rhino population plummeted. The first wildlife biologist to visit the area, the famed E. P. Gee in 1959, estimated that only 300 individuals were left at the time. The precipitous decline of rhinoceroses and tigers, just after Gee's survey, eventually led to the creation of Royal Chitwan National Park in 1973 and, two years later, to two other wildlife reserves—Royal Bardia National Park and Suklaphanta Wildlife Reserve—both former royal hunting reserves. In 1975, the British graduate student Andrew Laurie, the first to conduct a field study of greater one-horned rhinoceroses in Chitwan, estimated the total number of that species in the park to be between 250 and 280 animals.

~

When possible, biologists strive to conduct longitudinal studies, that is, research programs conducted over long enough time scales to yield insights that the typical field study, which averages one to

*Male greater one-horned rhinoceros (*Rhinoceros unicornis*)*

three years, often misses. In 1986, a Smithsonian Institution project that had been focusing on tigers since 1972 embarked on population monitoring of the rhinos, which has continued to this day with little interruption. These studies would allow us to track long-term trends in the population and, most of all, to determine whether it is rebounding and has responded to increased protection. So it was that, beginning in 1986, Vishnu, his fellow trackers, eighteen drivers, and I rode our elephants from one end of the park to the other to photograph every rhino we saw. Each day we would spread out with five elephants to search for a rhino. When we discovered a new individual or a familiar rhino that lacked a revealing set of photographs, we would surround the subject. The other four elephants would usher the rhino toward my sturdy mount, and my camera would capture its left side, right side, front, rear, and horn. I made a special effort to note irregularities in the skin folds, knobs of extra skin, cuts and scars, clipped tails, broken horns, odd pigmentation, and a variety of other characteristics that would aid in recognition of individuals. Over time, we amassed a rhino register with a distinctive name for each individual and personalized notes about where it lived, its sex, the size of its horn, and, for females, the presence and birth date of a calf.

After three years of surveying the park, we found that the entire Chitwan population of rhinos had grown from around 270 in 1975 to around 400 by 1988. The encouraging trend showed that even very large, slow-breeding mammals can begin to recover from bouts of near extinction when they are protected from poaching and a nucleus population of around 50 animals still exists. Rhinos are not bristly rabbits when it comes to breeding biology. Females reach sexual maturity at about six years, gestation takes sixteen months, and they give birth to a single calf on average every three to three and a half years. One key to a rapid rebound for any large-bodied rarity is to reduce the loss of adult females. Another is the ability of large herbivores to adjust a key aspect of their reproductive portfolio, what biologists call the interbirth interval. Normally, the period between the birth of one calf to a female rhino and her next birth is almost four years. But a species can cut that period almost in half if it enjoys excellent forage and plenty of space.

About 150 rhinos lived within a three-hour march of our field station. To search more distant haunts, we went on safari for two weeks at a time, an experience the trackers and even the elephants seemed to relish. Typically, we picked camping spots near water and good grazing for the elephants, deep in the heart of Chitwan's jungles. The safaris began in mid-February, with the approach of the hot season. The dried-up elephant grass blanketing the floodplains had been burnt to the ground in fires caused by lightning or set by elephant drivers to improve forage conditions for the wild grazers. Now a carpet of green shoots attracted herds of deer and hungry rhinos. Tigers, invisible for most of the year, stalked through the green blades and charred stems. For a two-month window, the invisible wildlife of the Chitwan grassland was just as watchable as wildlife in the Serengeti.

When our survey team reached the Narayani River, the western border of Chitwan, we had come to the park's wildest part. Unlike the animals that lived near our base camp, the rhinos here spooked easily, lacking much exposure to humans or domesticated elephants. They either ran away as soon as we approached or, in the case of some bulls, ran straight at us. Near the edge of the Narayani, the elephants spread out on the floodplain. The grass had yet to burn off here, so it still provided ample cover for the prey of the tiger—sambar deer, axis deer, hog deer, wild boars. Suddenly Mel Kali, bobbing through the thick grass to our left, banged her trunk against the ground with a loud thump: she'd picked up the scent of a tiger. "*Bhaag, bhaag!*" the driver whispered gleefully to me. "Tiger, tiger!" He gave the adventurous smile our mahouts flashed whenever danger was at hand. My elephant, Prem Kali, lifted her trunk over the rustling stalks to sniff the air and brought it down heavily on the ground. *Thump!* Then she emitted a deep rumble, picked up by the other elephants. The grass was so thick they couldn't see one another, but their subsonic vocalizations helped them communicate the movement of the predator.

The tiger, a big male, waited until we were nearly on top of him before he unleashed a thundering roar and flashed by us in a blur of orange and black. Unable to contain herself any longer, Prem Kali trumpeted, and the other elephants echoed her blast. In seconds the tiger was leagues of grass away.

By the end of the morning, our rhino photo shoot was over for the day and it was time to meet up for lunch. Kancha Bahadur Lama, another of Vishnu's childhood chums, was our jeep driver, and he planned to meet us at a crossroads to deliver a hot meal. Earlier, several families of fishermen—ethnic Bote people—had given us some smoked fish in return for scaring off the lurking tiger. Now we could look forward to our meal of fish and rice. We reached the rendezvous, but no Kancha Lama. No worries; faithful Kancha would show. The drivers removed the elephants' saddles and let them graze the nearby lemongrass. Soon the air was filled with perfume as the elephants' trunks plucked aromatic bundles of grass, whacked them against a foreleg to dislodge any dirt, and stuffed them into their mouths.

Thirty minutes, and then an hour, passed. Kancha Lama and our lunch still hadn't arrived. Vishnu immediately switched roles from head tracker to executive chef. Under his direction, the men dispersed and came back with the ingredients of a jungle tasting menu: *Grewia* berries, fresh wild ginger stems, the smoked fish, and his inspiration, peacock eggs Vishnu.

We sat down to our repast. The food may have lacked refinement, but it was infused with the good spirit of those who gathered and prepared it. *Grewia* fruits taste like wild blueberries and are a favorite of the sloth bear. This usually fierce, shaggy denizen of the floodplain dines on termites and ants, much as would an anteater or giant armadillo in other parts of the world. But when *Grewia* fruits are around, sloth bears feast on them as grizzlies do on wild huckleberries. The wild ginger stems were a revelation. Fragrant and with a bite, they could replace sorbet as a palate cleanser. Vishnu built a small fire, set the avocado-sized peacock eggs at the edge to cook in their shells, and warmed the smoked fish, which he had wrapped in lemongrass leaves. Lacking an egg timer, Vishnu waited for the first of the eggs to explode and then quickly removed the rest from the heat. The golf ball–sized yolks were quite edible, but the whites had assumed the texture of vulcanized rubber. I left my share for the jackals. An hour later, Kancha Lama finally arrived, red-faced and to much teasing, including how he missed out on the adventure.

The elephants had rested and were now bathing in the Narayani. With time on our hands, we lit up hand-rolled cigarettes known as bidis. Smoking by the riverbank, we watched busy cormorants hunt for their dinner. The bright green water shifted and swirled in eddies around sand bars. A few kilometers from where we sat, the river bent south on its way to meet the Ganges. In a month or so, the monsoon would arrive and the Narayani would flood, washing away all signs of our meal. But this perfect day—charging rhinos, a flashing tiger, fishing cormorants, the fragrance of elephants feeding on lemongrass, topped off by a turn as hunter-gatherers—would never fade from my memory.

~

Our census of the greater one-horned rhinoceros and data gathered from the thirty-five or so radio-collared rhinos we had been intensively monitoring led to a number of conclusions. The first confirmed the observation that Chitwan rhinos shunned the dense mature forest in favor of riverbanks where wild sugarcane swards grew. Much of Chitwan is composed of sal forest, the dominant forest type of the northern part of India and lowland Nepal. Wildlife biologist Bivash Pandav calls it "green desert." Sal is a member of the family Dipterocarpaceae, the most valued group of timber trees in Asia, comprising about 534 species and reaching its highest diversity in Borneo and Sumatra. Northern India and Nepal are covered by just one species, *Shorea robusta*; it grows tall and stout with deep-furrowed bark and straight, lightly branched trunks and remains green almost year-round. Despite the persistent greenery, few of the plants that grow in this forest, including the tannin-rich sal, produce leaves that herbivores like to eat. Our census confirmed that in much of the park, rhinos are relatively sparse because of the dominance of this forest type.

The rhinos also concentrated their feeding in a small area. One traditional rule in mammalian ecology, especially for herbivores, states that a species' home range is related to its body size. This is a critical insight in the study of rarity. Plant- and seed-eating rodents—a category that includes many narrow-range rarities—can still be abundant numerically. Their home range is often measured in square meters. At the other extreme, elephants and rhinos, according to the rule, should roam widely and live at low densities. Yet radio-collared female rhinos used only 3.5 square kilometers annually and males a slightly larger area, in contrast to elephants, whose home range might be as large as 30 square kilometers. The core part of the rhinos' range was actually less than one square kilometer, remarkable for a giant herbivore. Just as important, the home ranges of females, including those with calves, overlapped, so many rhinos were packed into a small area. That spring, the density in the riverine grasslands of Chitwan reached thirteen adults per square kilometer—among the highest densities ever recorded for a giant mammal. What, we wondered, accounted for this? And it wasn't just rhinos whose density was unusual, but the park's big carnivores as well. The home ranges of Chitwan's tigers and leopards are smaller than almost anywhere else, with male tigers averaging about 20 square kilometers and females around 6–10 square kilometers. For comparison, in the Russian Far East a male tiger's home range may be as large as 600 square kilometers.

Happily, our long hours of logging data on elephant-back were starting to answer this question, first for rhinos and then for tigers. The twenty-four-hour movements and feeding preference of habituated radio-collared rhinos revealed that not all grasslands were equal for them as feeding spots. The term "elephant grass" encompasses a host of extremely tall grass species, many of which are highly woody and unpalatable to rhinos once the plants grow past the shoot stage. Our results showed that rhinos prefer only about five or six species of these tall grasses. Tall grassland accounts for about 10 percent of Chitwan National Park's vegetation, and 90 percent of that 10 percent is dominated by two species that rhinos eat only as young shoots, one in the genus *Themeda*, the other in *Narenga*. The remaining sliver of grass habitat directly adjacent to the riverbed, covered by a species known in Nepali as *kans* (*Saccharum spontaneum*), or wild sugarcane, is what rhinos really prefer. The wild cane remains green for much of the year and sprouts anew in response to grazing, cutting, fire, or inundation. Year-round, kans accounts for at least half the monthly diet of the rhino; of all the rhinos' potential forage grasses, kans turned out to have the most protein and to be one of the most digestible. So rhinos are not only grassland specialists but also kans connoisseurs. Using their prehensile upper lips to wrap around the stems and their high-crowned molars to chew them up, rhinos are highly proficient mowing machines.

The kans grasslands and some of the other short grasslands adjacent to them or mixed in with them were also the preferred feeding areas for the tiger's prey. The density of deer and wild boars in the kans and riverine forests was among the highest recorded wherever this habitat occurs in Nepal and India.

Floodplains in Asia are among the most productive landscapes on Earth for rhinos, tigers, and other animals as well. Such rich, fertile soil supports, for example, the endangered swamp deer, extinct in Chitwan since 1950 but still present in small numbers in Bardia and abundant in the Suklaphanta reserve. Rare across much of its range in Nepal and India, the deer forms herds that number in the hundreds, until recently in the thousands, in open grasslands in Suklaphanta. The same is true for the endangered hog deer, which is rare elsewhere but quite common in the kans, sometimes traveling in herds of thirty or more. And wild water buffalo, which used to be common in Chitwan but now have been extirpated, were reportedly found only in the riverine grasslands. The foraging patterns of all these large-bodied species illustrate how the highly productive kans, the first elephant grass in the line of succession along riverbanks, is the keystone plant species offering the most critical habitat in this large-mammal ecosystem.

Another critical habitat element for rhinos is water. During the hot, steamy monsoon, rhinos are unable to sweat fast enough to cool off; to compensate, they become almost semiaquatic, spending up to eight hours a day submerged up to their nostrils. When they wallow on their side, their broad bodies form perfect sundecks for rows of amphibians. When the rhinos turn over, the frogs quickly shift to their new perch.

Rhinos concentrate close to rivers, near the ribbon of wild cane, and in areas pockmarked by wallows. Chitwan offers all of these features. With rhinoceroses so locally abundant there, I had to keep reminding myself just how globally rare these creatures truly are. On a typical hot spring morning in 1987 in the Pipariya grassland, for example, rhinos, especially mothers and calves, seemed to be everywhere. Within the span of two hours, in an area no bigger than a shopping-mall parking lot, we counted thirty-five individuals, the highest concentration I had ever seen. A morning stroll across the grassland would demonstrate the truth of local abundance to any ecologist who doubted that such a thing was possible, even for a rhino.

~

On a steamy July evening in Chitwan, we had just sat down to dinner when we heard tremendous crashing sounds in the forest next to camp. Galloping across the compound at full tilt were two male rhinos, one chasing the other. The bellow of the chaser sounded like the earth ripping apart. Ignoring us completely, the males thundered past the dining area and straight through a barbed-wire fence, snapping the strands as if they were party ribbon.

This vignette of competition among males sparked a new research focus for us: the link between dominance and breeding success in males. Male rhinos don't use their horns when they fight; instead they use daggerlike tusks housed in the lower jaw. Remarkably, we found, it was the size and condition of the males' tusks that determined who had sex and who was vanquished. When we recaptured losers to replace their radio collars damaged in the duel, the vanquished males always had shorter incisors or broken teeth, as compared with the intact, larger incisors of the males that usurped them.

This relationship between dominance among males and breeding success is vital to understand for both rarity theory and conservation, especially when rare species reach low population levels. It is of even greater concern where the species is polygamous and one long-lived male can often monopolize breeding for many years. This could in turn lead to inbreeding, which would reduce the genetic vigor of the population and lead to the kind of downward spiral that rare species must avoid. This mechanism is the presumed cause of severe declines among isolated small populations of cheetahs and some mountain sheep.

According to the data we collected, however, these rhinos were in no danger of genetic decline. By radio-collaring the seven bulls that had been dominant in our intensive study area during the five-year field project, we found that six different males had rotated through the top position. One reigned for as long as a year and a half, but the tenure of two other males lasted less than two months. By gauging the age of each calf they produced, we gained a proxy for how many rhinos each male had likely sired while dominant. In some cases, the answer was zero because females cycled into estrus, or breeding stage, only every sixteen months or so.

The rapid turnover among breeding males implied a high degree of genetic mixing, in our view. To test this theory, we took blood samples for future analysis from many of the rhinos we sedated. Ecologist Gary McCracken led this part of our effort and, during a visit to his lab in 1988, revealed some startling news. "You won't believe this. Your rhino samples have among the highest levels of genetic variation ever recorded for mammals!" The results ran contrary to a commonly accepted aspect of rarity, that rare species typically have little genetic variability, especially those on the brink of extinction.

How had the rhinos managed to accumulate so much genetic variability in the first place? And how had they been able to retain it when their numbers crashed after 1950? The answer to the first question was simple. As would most ancient mammals, rhinos over the course of millions of years had accumulated a lot of mutations—and thus genetic variability—during their evolution. When rhinos were common across their range, they were highly mobile and able to spread their genetic material among their populations and maintain a large breeding stock. In so doing, the species built up a large genetic reservoir. This is also likely true of other large mammals where individuals have moved long distances from their birth areas and there has been a lot of genetic exchange among populations.

The answer to the second question also seemed straightforward. We knew that the collapse of the rhino population was relatively recent. With more than 1,000 rhinos likely in the Chitwan area before 1940 or so, and a crash to 60 to 80 individuals by the early 1960s, only a dozen or so rhino generations had elapsed before the population started rapidly expanding again in the 1960s. McCracken explained: "Populations lose variability by rhino generation, not year by year—a rhino doesn't breed every year, and few rhino generations had elapsed since 1950. So little loss occurred." Here, then, was an example of a large-mammal population that had survived a bout of near extinction and still harbored high levels of genetic variability.

∾

When a large species becomes especially rare or extinct, there may be many consequences for the ecosystem of which it has been a part. Large mammals that dominate an ecosystem and then disappear or shrink to a level at which they become "functionally extinct" no longer perform their long-standing ecological roles, whatever those might be. In chapter 1, I described the experience that triggered my interest in rarity—understanding the role of greater one-horned rhinos in the dispersal of seeds of the tree *Trewia nudiflora* and the creation of *Trewia* woodlands in the grasslands. Lecturers in plant and animal ecology in the United States and Europe often mention the extinct terrestrial giants only in passing. Bison, mastodons, giant ground sloths, North American rhinos, woolly rhinos, and their allies no longer play their former parts as landscape engineers, so they are typically ignored except as historical curiosities or examples among the early extinctions. But through their browsing, trampling, grazing, wallowing, and manuring, giant herbivores have over time played a major role as nature's architects, shaping the evolution of plant traits.

In most of the world today, no living laboratories remain in which to test theories about the ecological roles of big mammals. In areas of Chitwan, however, giant herbivores lived at such high densities that it was possible to study how plants and giant mammals might have interacted since the Miocene epoch, over 20 million years ago.

All one needed was an elephant to maneuver through the grass and follow the rhinos, a bit of curiosity, and some basic gardening skills.

The sheer size of the latrines and the dense stands of *Trewia* trees they supported were a revelation of sorts. So was a paper I had read prior to coming to Nepal, which presented a controversial hypothesis formulated by Dan Janzen, one of ecology's leading thinkers, and Paul Martin, a world-renowned paleontologist. They dubbed their idea "the megafaunal fruit syndrome," and we speculated that it might explain the phenomenon of latrine groves. Janzen and Martin proposed that the now extinct New World tropical megafauna—ground sloths, elephant-like creatures, prehistoric horses, and other giants of the Pleistocene—once played a major role in the dispersal of the woody flora. They argued that the long coexistence of neotropical plants and large mammals that ate large fruits influenced the evolution of fruit and seed traits of some plants so they would be consumed and dispersed by giant mammals.

What is a megafaunal fruit? The short answer: picture an avocado or a mango. In theory, such a fruit would be eaten by big mammals with large mouthparts that could easily handle fruits too big to swallow and too hard for smaller contemporary fruit eaters—birds, bats, rodents, monkeys—to crack open. The classic example is a dull-colored large fruit that shields its big seeds in a thick shell or rind that is hard when the fruit is ripe. Presumably, coevolution occurred because giant mammals provided a better vehicle for dispersing the seeds of these plants than did small fruit eaters or gravity alone. Those plants that created fruits with features attractive to big mammals received more attention and better dispersal to safe germination sites. The mammals received a highly nutritious and concentrated meal from the pulp surrounding the seeds, which they digested before passing most of the seeds intact. Such efficiency in feeding for the big mammals and distant dispersal away from the parent tree helped to cement this relationship—an evolutionary pas de deux—between two seemingly unlikely partners. For the rhinos of Chitwan, *Trewia* fruit fit the bill as a megafaunal fruit. *Trewia* is the most common tree in the forests bordering the Rapti River, but no arboreal or flying creature touched its ripe offerings. Instead, these large green fruits, about the texture and size of young apples, carpeted the forest floor, and the rhinos vacuumed them up.

To test the hypothesis that the dominant plants that had evolved in Chitwan were those not only eaten but also dispersed by megaherbivores, I had to look at the situation with and without the big wild mammals to assess the strength of this interaction. In the absence of rhinos, cattle, deer, or water buffalo—the large fruit eaters—uneaten fruits pile up and rot, their seeds are attacked by bright red insect seed predators called largid bugs, and there is almost no germination. But what we had documented so far was the importance of large herbivores to fruit removal. We didn't know anything about seed dispersal.

To find answers, we needed a volunteer rhino to observe. Sunder Shrestha offered us Kali, the one-horned rhino at the National Zoo in Kathmandu. We drove up to Kathmandu on several occasions to conduct feeding trials. Kali was a willing participant, even though she had probably never tasted a *Trewia* fruit because she had been taken to Kathmandu as a calf. The fruits were placed on a ledge in her enclosure so she could accept or reject them. During the first trial, using her prehensile upper lip as a rake, she ingested all 114 of the fruits placed in front of her in 10 minutes. On a second trial a few weeks later, she ate 300 fruits in 50 minutes. Each time, we waited for the seeds to come out the other end.

Kali let us collect her dung each day, and Vishnu, Kancha Lama, and I painstakingly picked through hundreds and hundreds of kilograms of bright green rhino dung to extract the shiny black *Trewia* seeds. The first seeds appeared within 46 hours after ingestion, and the peak occurred between 60 and 84 hours after Kali ate them. The last whole seed to emerge took its grand time, appearing 172 hours after intake. Because we knew there are approximately 3.2 seeds per *Trewia* fruit, we estimated that about 60 percent of seeds ingested made it through Kali in the two feeding trials.

Being eaten by a rhino, then, was not a dead end for *Trewia* seeds, which is a good thing, given that during the fruiting season as much as 10 percent of the rhino's diet can be *Trewia* fruit. It also gave the seeds a brighter future. It turned out that gut treatment and manuring of seeds had no effect on seed germination. But being encased in giant pillows of dung gave young seedlings a huge boost over seeds planted in soil.

And where rhinos defecate really matters. Like the rain forest tree seedlings in the Peruvian Amazon, *Trewia* sprouts do poorly under the parent tree in the dense forest; they belong to the caste of seedlings that biologists call shade intolerant. To prosper, they need to be manured into grasslands exposed to intense sunlight. By planting seeds and seedlings under various conditions and moving latrines from shade to sun and vice versa, we were able to establish that rhinos are essential to the survival of *Trewia*. Clearly, the latrines were a significant phenomenon, not only as a landscape feature but also as a beachhead for woody plants in a sea of grass.

Giant herbivores shape their world in other ways as well. In the winter months, the rhinos in Chitwan, for example, shifted from grazing grasses to browsing tender shoots and leaves in the riparian forest next to the wild cane. During December, the sound of rhinos walking over saplings of wild avocado trees filled the damp night air. By the end of February, when the rhinos switched back to grazing, the riverine forest resembled a war zone of flattened wild avocado trees. Yet rather than dying off, the trees adapted to this abuse. They grew horizontally and sent out new shoots snaking in all directions, like a woody medusa.

What might a rhino-less forest look like? Given the decline of rhinos in so many places, this wasn't an idle question. We constructed stockades sturdy enough to exclude rhinos. After three years of protection, we compared tree growth inside the stockades with that in paired plots where rhinos had enjoyed free access. The results were striking, even to the uneducated eye. Where the wild avocados were protected, their stems grew straight to the sky;

where rhinos worked them over, trees remained stunted. Rhinos do the same to wild rosewood and a tree genus called *Mallotus*. In a similar way, elephants suppress silk cotton trees, Asia's analogue of the baobab. If left unattended in a silk cotton grove, elephants will girdle the trees to reach the tasty inner bark and kill them, rapidly converting a silk cotton tree savanna to a grassland.

So, in a forest still stocked by its Pleistocene herbivores, only the saplings that are unpalatable to the giant browsers will ever reach the canopy. For the invertebrate species, such as sucking bugs and beetles, that can eat the trees unpalatable to the large mammals, the more favorable ratio is a blessing. For other browsing herbivores that eat the same plants as rhinos do, the trampling behavior of the pachyderms leaves more browse potentially available in the layer they can reach. In short, an Asian forest minus rhinos and elephants will barely resemble one where they are the dominant browsers. Once-in-a-century monsoon deluges may reset the ecological clock on the floodplain. But between the flood events, it is truly the big beasts, through their manuring of seeds and intense pruning, that influence the structure and composition of the riverine forests and savannas.

Beyond Asia and its rhinos, rare vertebrates serving as ecosystem engineers are evident in many environments. Besides the fruit-eating birds of paradise in New Guinea and hornbills in Asia, the large frugivorous birds in Amazonian forests and fruit-eating monkeys have an important impact, as we saw in Peru. In the habitat of the Kirtland's warbler, beavers, even at low abundance, create habitat with their damming of streams. Studies suggest that prehistoric mammals, such as woolly rhinos, created the productive Pleistocene high-latitude steppe through their intense grazing and trampling, just as the greater one-horned rhino and white rhino maintain what one biologist has termed "grazing lawns," where the close cropping of grassy areas by giant herbivores with prehensile lips keeps a low sward at a height that benefits many smaller herbivores in their search for nutritious grasses. In the boreal forest zone, the giant today is the moose, but this browser eats mostly twigs. How different would the boreal forests look if they still had their full complement of mammoths and rhinos?

~

Although rhino numbers were increasing in Chitwan while we were there, Hemanta Mishra knew that an epidemic, a severe outbreak of poaching, or a catastrophic flood could decimate Nepal's only population. In the 1980s, translocations of endangered rare mammals to reestablish extirpated populations had become a centerpiece of conservation efforts around the globe. The more populations were spread across their historical range, the theory goes, the greater was the likelihood for the species to adapt to new or changing conditions. Creating a second and even a third rhino population within their former range could be an important hedge against local extinction. As a rule of thumb, conservation biologists advocate that at least ten populations of an endangered mammal species be established initially, with at least 100 founders in each, to reduce the threat of extinction. More than two hundred years ago, rhinos roamed across the Bardia and Suklaphanta reserves in the far western part of Nepal's Terai lowlands. Now these areas were well protected and ready to accommodate their former residents.

So, in the early evening of a winter night in 1986, I found myself riding shotgun in one of three timber lorries the government had provided for transport. The payloads were three tranquilized but awake rhinos housed in giant wooden crates. Captured earlier in the day, they were bound for Bardia, 250 kilometers to the west. Under the full moon, our caravan rumbled cautiously along the winding roads of southern Nepal. I sat in the cab listening to Vishnu's stories, which took my mind off how risky this venture seemed every time our rhino passenger shifted in her crate.

After fifteen hours, one flat tire, and one truck temporarily stuck in a river, we made it to the release site in Royal Bardia National Park. By now, Vishnu and his team had perfected the capture-and-release protocol. The door of the first crate was opened, and out rushed the first female rhino. The next animal to be released, a subadult male, burst out looking for something in his path to crush. The last rhino to be unleashed also seemed ready to charge but changed her mind and walked into the tall grass.

Ironically, attempts to translocate and recover some of the mammals smaller than rhinos proved much more difficult. Blackbucks— a beautiful antelope species—moved from a wild animal park setting into protected areas thrived at first but then disappeared if the grass grew too tall and gave cover to hunting tigers. The smaller, more highly strung deer and antelope also had high mortality during capture and translocation. No one could capture bristly rabbits. Although pygmy hogs have recently been captive bred and released back into the wild in India to create new populations, captive breeding of musk deer had stalled because of significant mortality during the process. For some reason, the greater one-horned rhinos responded beautifully to the morphine derivative used to sedate them. Could this be one more aspect of their rugged nature? Perhaps those same selection pressures for resistance and resilience made them, however inadvertently, easier subjects to ship out for restoration.

Those first three translocated rhinos were eventually joined by ten others. In late December, seven days after starting the capture operation, we left Bardia, having deposited the last of the group of thirteen. When we captured the rhinos for this translocation, I had touched the horns of each. I wondered how the fate of the five rhino species might have differed had they never evolved facial horns. Would their size and habitat specialization still have rendered them so endangered? The absence of a horn, or even the drastic measure of dehorning rhinos—tried in Namibia and Zimbabwe to deter poachers—would have little effect on a major threat to their survival posed by habitat loss, as is the case today. On the way back to Chitwan, we drove through one poor mud-walled village after another. The rice harvest had just been taken in, and stacks of straw lined the way. In this region, with one of the world's highest birthrates, pressures on land and critical rhino habitat could only escalate.

We pulled up to a teahouse not far from Lumbini. The stall was swarming with flies, thousands upon thousands of them, coating the hanging sweets and landing on the rims of the unwashed teacups. My desire for a snack quickly diminished. The Zen phrase "Where there are men, there are flies, and also Buddhas" came to mind. In truth, we were near the birthplace of Gautama Buddha, who as young King Siddhartha gave up his royal life to wander as a peasant and find enlightenment. Lumbini has become an international draw for devout Buddhists. Just as Nepal's religious shrines draw wide support and popular praise, I thought, so should its miraculous recovery of endangered wildlife, against all odds, in one of the poorest nations on Earth. I left my daydream as Vishnu gestured to me. The ever-efficient tracker had located a cleaner tea stall, and we sat down for a refresher.

What started as a small research program in 1986 had grown by 1999 into one of the world's most successful restoration efforts of a rare mammal, winning international acclaim for the country and for Mishra. By the year 2000, more than 85 rhinos had been translocated to two different populations in Bardia and 6 to Suklaphanta. The combined rhino populations in Chitwan, Bardia, and Suklaphanta had shot up to more than 600. Importantly, the local people also benefited. For many years Chitwan had been the major wildlife tourist destination for visitors to Asia from industrialized nations. The chance to see rhinos and possibly a tiger drew these nature lovers. New legislation was passed to benefit local communities: revenues generated by park entry fees and hotel concessions were recycled into village economies on the periphery of the park.

~

I wish the story could have ended here—an account of a triumphant recovery of a once common species made rare by humans but given a second chance. But Nepal soon plunged into its darkest period in modern history. After a popular call for a democratic government, King Birendra accepted the change from an absolute to a constitutional monarchy in 1990. But democracy failed to take root, and word began to filter out of the western hills of a violent separatist group that called itself Maoist and sought the overthrow of the monarchy and the creation of a communist state.

Then, in 2000, the unthinkable occurred: most of the royal family was assassinated by one of its own. The kingdom went into mourning as the middle brother, Gyanendra, assumed the throne. Units of the Royal Nepalese Army that had been stationed in the national parks, partly to protect rhinos and other wildlife, were soon shifted into the hills to fight the Maoists, leaving the parks wide open to emboldened poachers. The devastating results provided a glimpse of how civil unrest or even civil war can be the death knell for rarities.

In Bardia, all of the 70 or so rhinos that had been translocated to the Babai Valley were wiped out. Chitwan suffered a staggering blow, losing about 170 rhinos to poachers. Its rhino population dropped from around 550 to about 370, about the same level as when we finished our census in 1987. For a while, all seemed lost.

And then, in 2005, Nepal began to right itself. The citizenry rallied, a peace movement emerged, and the government negotiated a cease-fire with the Maoists.

When the smoke cleared, our survey teams returned to assess the damage in Bardia and Suklaphanta. The Babai population of rhinos was gone, but all was not lost. The first group of rhinos moved to the western border of Bardia along the Karnali River in 1986 had grown from 13 to more than 30 animals. Several calves had been born in Suklaphanta, where the rhino population had almost doubled from the 4 reintroduced there. Across the border in India, rhinos sent from Chitwan had helped start a founder population in Dudhwa National Park that remained stable. A few colonists from Bardia had crossed into India's Katarniaghat Wildlife Sanctuary and started another breeding group, and then some from Suklaphanta had crossed into India, to the Kishanpur Wildlife Sanctuary. So while overall numbers were down, the risk of extinction—and the promise of recovery—had been distributed among five reserves.

There was no way to deny that the large number of rhinos poached in Chitwan and Bardia represented a enormous loss. But when a few more were killed by poachers in 2006, the people of Chitwan District took action. They gathered 100,000 signatures on a petition demanding that the government uphold its obligation to protect the rhinos. This remarkable display of popular conservation support is a rarity itself. It demonstrated that even in the poorest regions on Earth local residents may recognize their unique heritage and have pride enough to want to preserve it. There were also good economic reasons for doing so. By 2009, as much as $400,000 per year was going to park border communities to build schools, health posts, and roads. More rhinos meant more tourists, more tourists meant more revenue for local development and protection, and that completed the circle for more rhinos.

Now protection has returned to the Terai reserves, and the rhinos are poised for yet another comeback. The census in 2011 revealed a total of 534 individuals, a gain of more than 100 animals in three years. So even relatively slow-breeding giant mammals can be quite resilient, capable of rebounding quickly, if we provide basic protection. But when protection fails, or poachers run rampant, the gains can be wiped out overnight. The poaching epidemic currently raging in South Africa, home to 80 percent of the world's rhinos, tells us that rhino poachers have become more organized, better financed, better armed, and more ruthless. In 2011, more than 450 rhinos were poached in South Africa; in 2012, the number might reach 600. The poachers are no longer poor villagers but now are operatives in international crime rings.

My experience in the elephant grass taught me that in the absence of our species, these giant mammals might be far from rare in the highly specialized habitat they prefer. That habitat was once abundant, and the rhino beautifully adapted to it and the sweep of the monsoon floods that so characterize its environment. Unlike the Kirtland's warbler and other habitat specialists, the greater one-horned rhino is heavily persecuted for its body parts and comes into conflict with villagers, and without protection its stomping grounds would soon be usurped by poor farmers looking for this last available piece of rice paddy. Yet, despite all these constraints, the species thrives if given a bit of protection. In fact, the same is true for virtually all large mammals, even those that reach maturity after a long adolescence and that have only one offspring at a time, long gestations, and multiyear interbirth intervals. Whether they are habitat specialists, like the greater one-horned rhino, or habitat generalists, like the African elephant, one aspect rises above all others in a shared conservation strategy: protect the females and let them breed and live to an old age, and recovery happens much faster than even the cynics could ever imagine.

We must up the ante if we are to recover these tough creatures, who would just as soon trample those who try to save them as those who poach them for their horn. Beyond rhinos, there are the stalking tigers, the grizzlies digging for tubers, the lions lounging under the acacias—all of these species remain unaware of our efforts on their behalf and sometimes see us more as prey than protectors. It is yet another sign of the cultural evolution of our own species that we can reconcile the conservation of species that we fear most and create sanctuaries where they can thrive.

There is a spiritual dimension to rarity, too, I believe, hiding there in the elephant grass, in the acacia savannas, or in the grizzly range. The presence of large, potentially dangerous mammals connects us to something deep and primal and teaches us humility in a way that is unique and precious. We must not lose it. Wild species that leave footprints larger than our own are now among the rarest of all mammals. Places where they still dominate the landscape must be part of the legacy we bequeath to future generations.

Chapter 6

Scent of an Anteater

HUMAN ACTIVITIES, AS THE STORIES of the greater one-horned rhinoceros and other species have shown us, sometimes bring species back from the brink of extinction. But more often they exacerbate rarity even to the point of disappearance, drive into rarity species once common, or further constrain those species that normally have narrow ranges or live at low densities. The most dramatic change happening today that is pushing already uncommon species toward even greater rarity is the conversion of rain forests and natural savannas into commodities production for industrialized agriculture. Big Ag, as it is now known, is largely mechanized, highly profitable, and controlled by multinational corporations. Some biologists and geographers describe extension of this trend as the future; increasingly, we live on a cultivated planet. The loss of natural habitats through nonagricultural use—that is, human settlements—and in nontropical areas is also high, but the conversion is greatest in the tropics and through big agriculture.

Few field biologists bother to check the daily price of soybeans or palm oil. This is an oversight because the market value of these commodities—along with that of beef, corn, sugar, and coffee—may over the coming decades define the future of rare species more profoundly than will any other driver of habitat loss. At present, nowhere is the conversion and fracturing of rain forests by industrialized agriculture in the world's hotbeds of rarity more evident than in Southeast Asia and Brazil. In Kalimantan and Sumatra, Indonesia, expansion of oil palm and wood pulp plantations threatens the most species-rich rain forests in the world. In Brazil, vast areas of the Amazon are turning into cattle ranches and soybean farms. In addition to causing habitat loss, such rampant conversion imperils climate stability. Nearly 70 percent of the greenhouse gas emissions released annually from tropical forests originate from agriculture-driven forest conversion in just two places, Riau Province, Sumatra, and the state of Mato Grosso, at the edge of the Amazon in Brazil.

The plight of the Brazilian Amazon grabs headlines, but the status of its neighbor, the Cerrado, remains veiled in obscurity. This omission is a pity because the savannas of South America hold the key to reaching a balance between safeguarding rarities and growing the food we eat. Few environmental journalists are familiar with the Cerrado, which represents about 21 percent of Brazil and where both the total amount and annual rate of habitat conversion is higher than in the Amazon region. Over the past fifty years, more than 55 percent of the native habitat of the Cerrado has been cleared to make way for crops and livestock. Only about 2 percent of the region receives formal protection from the federal government, and Brazil's Forest Code, at least on paper, requires protection of habitat on 20–30 percent (depending on the Brazilian state) of private lands.

The Cerrado borders the Amazon rain forest to the west and the green ribbon of the Atlantic Forest to the east. To the south lies the vast seasonal swamp known as the Pantanal. The Cerrado ranks as the world's most diverse tropical savanna, even richer than the miombo, a similar habitat in southern Africa. The miombo's infertile soils and tsetse fly infestations repel agriculturalists, whereas the Cerrado can be farmed for commercial crops after some soil modification. It has become the world's largest producer of soybeans and beef and soon will be a major producer of sugarcane. The Cerrado has the dubious distinction—along with the previously mentioned Indonesian provinces of Sumatra and Kalimantan—of being among the most biologically diverse landscapes being converted most rapidly to agriculture.

This endangered tropical savanna features an unusual trio of rare mammals—the giant anteater, giant armadillo, and maned wolf. Very few tourists travel to Brazil explicitly to see them, even though the Cerrado offers the best chances of a sighting anywhere in the world. The ecotourism value of these species per hectare is far below the return that ranchers receive for beef cattle, soy, and sugarcane. So how do these rare species, along with jaguars, pumas, tapirs, and other wide-ranging Cerrado vertebrates, cope with massive land-use change driven by human economics?

Biologists who point out that a number of species can coexist in environmentally friendly cultivated zones have coined the term "countryside biogeography" ("matrix conservation" in Europe) to characterize the study of this phenomenon. This new discipline is essentially the study of which species persist in agricultural landscapes, assuming that interspersed with intensively used farmland are patches of natural habitat. To explore this issue and its relevance to the preservation of rarity, our next stops include Serra da Canastra and Emas National Parks in Brazil, at the edge of an expanding agricultural frontier that threatens to plow under rarity. Here, biologists are using a startling field technique that, along with the global positioning system collars worn by jaguars and pumas in the Peruvian Amazon (see chapter 3), could revolutionize the study of rare vertebrates.

Map of Brazil and environs

~

"Look for an overripe, black banana moving through the grass."
Edson Endrigo, our nature guide extraordinaire, was explaining his
technique for spotting giant anteaters in Serra da Canastra National Park, just one of the rarities in this area. Obediently looking
up on the hillside, I spotted a two-meter-long mobile banana. We
jumped out of the van and circled behind a female anteater with

a baby clinging to her back. My two companions, David Wilcove and John Morrison, and I closely tracked her progress.

If the greater one-horned rhino seems odd and prehistoric, the giant anteater offers good company as one of the most peculiar-looking mammals on the planet. Both are ranked as threatened on the IUCN Red List of Threatened Species. The *tamanduá-bandeira*, or *papa-formigas*, as it is known in Brazil, cuts a comical figure, sporting an elongate, arching snout and bowlegged limbs, all ending in an immense shaggy tail. The rest of the body is shaggy, too, featuring a striking long pelage of dark bands on light. The female in front of us moved along like an animated throw rug.

An anteater walks on thickened pads on the outsides of its paws, as its digits are turned under its feet. An observer might think of this awkward creature, with its poor eyesight, bad hearing, and odd gait, as defenseless against secretive jaguars and pumas. That would be a miscalculation. With its acute sense of smell, the anteater can make up for its nearsightedness. If cornered, it will stand up on its hind legs and slash with its massive claws any human or feline predator foolish enough to tangle with it.

The mama anteater stopped and flicked her tongue in the dirt. Unlike the vast majority of mammals, the giant anteater lacks teeth. It has no real need for them because it inserts its long, narrow tongue into crevices, removes ants and termites with its sticky saliva, and swallows them whole. Crouching downwind, I inhaled deeply to catch its scent and wondered if consuming 30,000 ants a day gives this creature, or its flatulence, the odor of formic acid. I smelled nothing unusual.

Edson shooed us back into the van to pursue the other goal of this outing: to search for the Cerrado's rare endemic birds. Earlier that morning, he had led us to the Brazilian merganser, an incredibly rare duck that, like the Kirtland's warbler and the greater one-horned rhinoceros, is an extreme habitat specialist, one that lives only on the fast-moving, clear streams of the upland Cerrado. Contamination from gold mining (now banned) caused the decline of

*Giant anteater (*Myrmecophaga tridactyla*)*

this species. Edson guided us down a canyon to give us a fabulous view of a bird whose entire global population was probably no more than 250 individuals. Mergansers are elegant-looking ducks, but the Brazilian version has a startling profile, accented by its pointy head feathers. This solitary female had chicks perched on her back as she guarded them through their first week of life.

As we moved on, nothing could have prepared us for the next sighting. In a grassland to our left, Edson's sharp eyes spotted a hovering bird. It was the most dashing of raptors, an aplomado falcon. The aerial predator was preoccupied, following something gliding through the tall grass. Then a head with pointed ears emerged. The maned wolf looked around for a second and moved on. The falcon persisted, perhaps planning to feast on the large insects or birds scared up during the terrestrial predator's afternoon hunt.

Shadows fell over the rugged escarpment in the distance as the afternoon wore on. We drove out of the park and entered the agricultural zone—"the Ag," for short. An hour later, our magical sightings of the wolf and the falcon, the merganser and giant anteater, began to feel like a dream. This had been only a first taste of the Cerrado's wildlife. To learn more about how these animals navigated the last natural pockets embedded in a landscape of soy and cattle would require a longer stay, and for that I had decided to join an unlikely pair of long-term researchers.

March 2008. The sea of grasses undulated in the warm, dry breezes. A tall, blond woman dressed in khakis and field vest reached down to release her dog from its leash. "Okay, Mason, let's go to work!" The dog dashed into the tall grass of Emas National Park. Every so often a grassy wave broke over the upright tail of the black Labrador retriever as he bounded through a large marshy area bordering a palm glade. The tail zigged and zagged through the wet pampa. Carly Vynne, then a PhD candidate at the University of Washington in Seattle, kept her eye on the dog. Within minutes, Mason returned with a look of great urgency. "What is it, Mason? Let's go look." Having grabbed her attention, Mason led us back through the muck and stood with his nose pointed toward the base of a grass clump.

At first we couldn't see anything. Then we bent down and noticed a cylindrical dropping half submerged below the tussock. Bingo. In a vast expanse of grassland filled with thousands of smells, Mason had detected the scent of rarity: he had sniffed out the droppings of a giant anteater. Carly could barely contain her excitement. Gathering herself, she placed a sample of anteater dung in a vial of preservative to protect the scat sample and the precious strands of DNA it contained from further degradation. Those convinced that dogs are superior to humans praise their loyalty, good nature, and capacity for unconditional love. Scientists appreciate another canine advantage—dogs have an uncanny sense of smell, surpassed only by that of bears and, by coincidence, the giant anteater. The homely bloodhound, with the keenest nose of any dog, possesses a sense of smell 300 times more acute than that of its handlers. Bloodhounds can detect a scent nearly two weeks old, but they are a lot harder to train than Labradors.

Sniffer dogs have recently been recruited for biological field studies because they excel at locating the fecal tidings of rare mammals. Carly had invited me to join the last year of her fieldwork in her study of a group of rare South American mammals in the Cerrado. The maned wolf, jaguar, puma, giant anteater, and giant armadillo of the continent's pampas and central savannas are vestiges of a rich Serengeti-like fauna that flourished in the Pleistocene epoch, 15,000 years ago. Today, their secretive behavior, low population densities, and ability to hide in the waist-high grass make sightings of these charismatic vertebrates quite rare. Their presence in the agricultural landscape remained an open question. The small size of the existing Cerrado parks and the wide-ranging nature of these species probably meant that some of them lurked out there in the ranchlands as well.

Field biologists who study the habitat use of rare mammals look for any sign: a scrape, a footprint, or the unexpectedly precious gift, a dropping. Miraculous advances in molecular biology have enabled researchers to extract strands of DNA and hormones from animal droppings, transforming the lowly fecal deposit into a gold mine of information. A scat sample can reveal the species of the depositor, individual identity, sex, reproductive status, diet, and health. Moreover, accumulated droppings from any single species, giant anteater or jaguar, yield the most prized data of all for rarities—density, home range, and population size.

The challenge, of course, is to find the fresh material from secretive animals that are often solitary and only part-time abovegrounders, as is the giant armadillo. The first three years of Carly's study, and other studies like it, had begun to show such promise that by 2010 human-dog research teams had gone global. Scent dogs are now used to study grizzly bears, Mexican wolves, wolverines, fishers, Javan rhinos, Indochinese tigers, Amur tigers, and other secretive mammals. The roots of this booming human-canid collaboration, however, trace back to an animal shelter outside Seattle, Washington.

~

If popular belief grants cats nine lives, dogs most certainly deserve at least two or three. Heath Smith told me that he had no opinion on animal karma when he entered the local shelter in Enumclaw, Washington, on a gray day in March 2004. He was not on a mercy expedition. Once inside the pound, Smith began bouncing a tennis ball off the concrete floor. Some dogs wagged their tails and rose to lick the visitor's hand but ignored the ball. A long-legged black Labrador pressed against the edge of his cage. He ignored the stranger but fixated on the delightful object Smith tossed in the air. The dog trembled with excitement and panted heavily. Here on display was precisely the kind of overwrought behavior bound to discourage even bighearted adopters. The shelter employee offered some details. "He was picked up running along the highway; must have gotten lost while hunting. You sure that's the one you want?" Heath nodded after taking a minute to confer with his boss by cell phone. "Great," answered the relieved keeper, "because he had twenty-four hours left."

Outside the pound, Heath escorted Mason into the cab of his pickup for the drive home. He turned onto I-90 and made for the house of his employer-landlord, Sam Wasser, a renowned conservation biologist at the University of Washington in Seattle. Heath was an accomplished dog handler, and Mason was one of the first candidates to become a new breed of working canine. The dog seemed delighted to be out of the kennel and most interested in the tennis ball hidden in the pocket of the person he was now accompanying. So began an improbable journey, from incarcerated stray on death row to professional scent dog set loose in the Brazilian outback.

Carly was one of Sam's students. She decided to try out scent dogs in her pilot study site, Emas National Park, in south-central Brazil, home to all of the Cerrado's largest vertebrates. "Ema" is the Brazilian name for the greater rhea, an ostrich relative that is common there. So are many other species that could be dangerous to a domestic animal bounding across the wild Cerrado. Having been trained to heel was vital if a dog were to stumble upon white-lipped peccaries, for example, which could tear a dog to shreds. The Cerrado is also home to fifty-three species of snakes. Carly wore snake guards to protect herself from the poisonous ones. Vipers, rattlesnakes, and fer-de-lances are all capable of killing a canine in one deadly strike. She was worried about unexpected encounters with anacondas, too. So she carried a machete in case there was a confrontation.

Protected areas are the best places to sniff out giant anteaters and giant armadillos in the Cerrado. Most grassland and dry forest mammals of South America are essentially rain forest species that have adapted to residing in seasonal forests and savannas. Thus, their extinction in the Cerrado may not spell the end of the species because others of their kind may still lurk in numbers in the forest. The maned wolf is an exception. Biologists call it an "obligate" grassland species because it cannot survive in the forest, having adapted, like the lion, to hunt its prey in open country.

Parks such as Emas, which is 1,320 square kilometers in size, are established to protect rare species, and they often perform well for habitat specialists and global rarities that live at high densities in small areas. The wandering kind of rarities, such as the maned wolf and giant anteater, that live at low densities can be much harder to protect. Outside of the deep Amazon, their world is changing by the minute. Circumstantial evidence shows that many rare tropical carnivores disappear when they leave the safe confines of their reserves.

The purpose of Carly's study was to determine how wide-ranging, low-density species such as maned wolves, jaguars, and pumas navigate the countryside and to learn how they circulate among the highly altered and fragmented habitats outside reserves. She also wanted to learn whether these wild species can shift from living in natural grasslands and adapt to feeding in soy fields and cattle ranches. Finally, using maned wolves as a test case, she wanted to learn if animals that live in the Ag are more stressed, less healthy, or less reproductively active than individuals that live inside the park. New techniques that Sam and Carly were developing back at the lab allowed them to measure health from hormones extracted from the dung. If Carly and Mason could find enough scats of this grassland wolf, this charismatic species might be the most revealing one to study.

Biologists use the terms "source" and "sink" to define the dynamic at work here in the Cerrado. Source sites are places where recruitment, or population increase, exceeds mortality. The expectation is that maned wolves have greater breeding success inside Emas than in Ag land, but because the park is too small to hold all the individuals born in the reserve, breeding individuals must spill out into the surrounding landscape. Sinks present the opposite situation, in which maned wolves (or members of another species) emigrate from the source to a location where they die in higher numbers than are recruited into the local population. With not enough large, preferably linked sources, and too many sinks, the species eventually will die off.

Some biologists avoid following their species into farmland or altered habitats. But in order to answer her questions, Carly randomly selected sampling routes that wound through the park as well as an additional 4,000 square kilometers of soybean fields, cattle ranches, and forest fragments on adjacent private lands. Curious about this woman and her dog, the local ranchers agreed to allow her to roam freely.

Within weeks after arriving in Emas, Mason had made the Cerrado his home, adapting well from a stint in the bitter cold of the

Canadian Rockies, where he had worked on wolf, caribou, and moose scats. Now he was in the dry oven of the Cerrado and was mastering Cerrado mammal spoor: puma, jaguar, giant anteater, giant armadillo, and maned wolf. To train Mason on tropical mammals, Carly had obtained sample scats from zoos in the United States and from Leandro Silveira, the dean of Brazilian carnivore biologists, and his biologist spouse, Anah Tereza de Almeida Jácomo, whose home base was Emas.

When Carly told her friends about her project, they all expressed admiration for the maned wolf. The maned wolf, though, it should be said, is not actually a wolf at all, bearing no relation to the gray wolf. The vivid red fur, trimmed in black and white, and the lovely mane of this Cerrado carnivore make it one of the most striking of mammals. The showstopper is to watch a maned wolf's aristocratic gait—a smooth trot on its strikingly long legs that matches the elegance of its coat. Nothing else on nature's runway compares to this handsome wild canid.

If the maned wolf is not a wolf, and the giant anteater an evolutionary oddity, no mammal is as strange as a giant armadillo. Its armorlike plating and bullet-train shape make it the perfect inspiration for a futuristic subterranean vehicle. Its huge claws, designed for digging its way through the underworld, are also impressive. Few biologists had ever studied this species before, and those who did mostly focused on its burrowing behavior and diet. Few had even found one alive, with or without the help of scent dogs. George Powell, our jaguar tracker from Peru (chapter 3), told me that once his research staff had heard loud snoring sounds coming from a burrow. They crawled in and pulled out a sleeping giant armadillo. Leandro Silveira also had an anecdotal encounter. But that was the extent of giant armadillo natural history—a few paragraphs.

These three rare mammals can all be described as unusual looking. So one might wonder if odd body plan, at least to human observers, is a predictor or correlate of rarity. I asked Carly about the link between countenance and ecology of giant armadillos, giant

*Maned wolf (*Chrysocyon brachyurus*) being followed by an aplomado falcon (* Falco femoralis*)*

anteaters, and maned wolves and what it might tell us about rarity. Her answers had as much to do with the energetic balance of these mammals as with their appearance.

"Even though these creatures eat different things, they specialize on food items that just don't support high densities of big mammals," she responded. This, not appearance, is key. Giant anteaters and giant armadillos are equipped to feed on ants and termites. While their food resource is ubiquitous, it is also of low nutritional quality (with some exceptions, such as fat-filled winged termites). The physiological and behavioral adaptations of these animals to their food resources may account for both their looks and their rarity in nature.

Giant anteaters have low metabolic rates relative to their body size. One consequence of their slo-mo lifestyle is that they produce only a single offspring at a time and only every other year. They occur at their highest-known densities in grasslands where their food resources are concentrated. But even here, anteater populations are severely constrained by the wildfires that regularly burn through. So even where their favorite foods—ants and termites—are highly concentrated, giant anteater populations are often knocked back by killer fires racing across the pampas.

Maned wolves, which are restricted to the grasslands of central South America, also occur at naturally low densities, half of our two-part condition of rarity. The maned wolf is the largest canid that does not hunt prey larger than itself. Its body mass may well be the limit at which a canid is able to survive on small prey—primarily rodents, birds, and armadillos, heavily supplemented with fruit. To meet their dietary demands, maned wolves traditionally forage across large home ranges of about 70 square kilometers, thus contributing to their natural rarity.

These unique adaptations—the elongated, toothless skull of the giant anteater that accommodates its extra-long ant-lapping tongue; the large claws of the burrowing giant armadillo, used for ripping open termite mounds; the fox-on-stilts appearance of the maned wolf—can be viewed in a new light. These striking features are all the evolutionary products of highly specialized feeding behaviors. Might they be unfavorable attributes in a changing world? An intense specialization on a highly patchy food source, such as termite mounds, works only if the species in question—in this case an anteater—can move effectively between patches. Thus, if the species is to thrive here in the Cerrado, surrounding ranchers would need to keep the termite mounds in their cattle pastures and anteaters would need corridors to reach these patches of rich termite concentrations. The long legs of the maned wolf might allow it to travel long distances easily and pounce effectively on its abundant prey. Being a medium-sized predator but basically subsisting on abundant small mammals such as rats means that you could find your principal food source in any open habitat. Natural selection may have no foresight, but it seems to have left the maned wolf a better chance of survival than the other members of the Cerrado trio.

Before heading out in the morning, Carly checked her data sheets and GPS unit and strapped on her snake guards. She counted the bottles of drinking water on hand for us and Mason, and we set out. When we reached her starting point on the transect, Carly released Mason with the command "Let's go to work!" Nose and tail in the air—the posture of a skilled scent dog—Mason trotted off into the grass.

The scent dog weaved back and forth across the route, off leash but always within sight of Carly. Within minutes, he came running back to fetch us. Then he raced back into a grassy area under some trees and sat down with his nose pointing a few inches away from a pile of maned wolf dung. Despite being an excellent fieldworker, Carly admitted she would have walked right past this scat had she been on her own. She immediately praised the dog and set about collecting the scat and noting its location on her GPS unit. The Cerrado project marked only the second time scent dogs had been called to duty outside the United States and Canada and the first under the tropical sun. By the end of a six-week pilot study here in 2004, three dog-handler teams had collected more than 650 scat samples from pumas, jaguars, and maned wolves. Their initial success made them believe that this novel technique could work well in the hot Cerrado.

We stopped for a break so Carly could give Mason a drink of water. The bells on his collar jingled as he lapped up the water from his bowl. The bells were a holdover from his grizzly work, a safety precaution designed to warn grizzlies that the team and Mason were near. Here in the Cerrado, Mason had wakened sleeping giant anteaters, and the nearsighted creatures merely ignored him and moved on as Mason retreated. He had also roused tapirs dozing in the grasslands, and Mason gave them a wide berth. Once he came face-to-face with a coiled rattlesnake, but Carly gently called him back to her side and the curious dog left the snake alone.

Here, the main reason for the bells was to warn off herds of peccaries. Only two months earlier, Mason had had a potentially fatal encounter. He was plowing through the grass and ran into a large gang of fierce peccaries. One turned and attacked, slashing Mason across his rump. Fortunately, Carly was close by and rushed Mason to the vet. After a few weeks of rest, Mason was ready to return to action, more wary of peccaries than ever before.

When we returned to camp, Carly took out Mason's bowl and fed him his ration. Rather than put him back in his holding crate for his afternoon siesta, she left him on the porch, tethered to a post by a long chain. Out of the nearby forest came a female black-and-white curassow. This heavy-bodied, sharp-clawed ground dweller was a favorite of local hunters but not a bird to mess with. She made her way over to where Mason was resting, but rather than claw at him as a possible predator, she snuggled close. According to Carly, this had become a daily routine. The male black-and-white curassow is all black, like Mason, and she may have seen him as a larger version of a possible mate. A female curassow in love, but one with a mean streak. This same female had run down one of Leandro's roosters and killed it.

The next morning we left Mason and his curassow flame snuggling on the porch while we toured Emas by car. We drove a long way to the northern border of the park but saw no anteaters or rheas. Their absence surprised me because we came upon huge numbers of termite mounds sticking up in odd funnel-like shapes. The mystery of the missing anteaters had a gruesome explanation. A catastrophic grassland fire had swept across Emas in 2005. The long fur of the anteaters had turned them into panicked torches, and five years later the population had yet to recover, in part because of their slow breeding pattern.

A central tenet of park design is to create reserves large enough to allow wildlife populations and natural processes, such as fire, to fluctuate naturally, with little or no human intervention. In this case, Emas would need to be several times larger than the area burned by the worst grassland fire of the century. An alternative design scenario would require that Emas be well connected by habitat corridors to other anteater reserves, to maintain a resupply route if a population inside a reserve is decimated by fire, poaching, or disease.

Somewhat discouraged by the lack of wildlife, we headed for lunch at a canteen attached to a run-down pool hall on the park's outskirts. Over a meal of rice, *plátanos*, and beefsteak, I mentioned that my Serra da Canastra birding trip, with its added sightings of maned wolves and giant anteaters, had sparked the collective curiosity of the van passengers. Together with Wes Sechrest, then head of the Global Mammal Assessment project of the International Union for Conservation of Nature and Natural Resources, John Morrison, David Wilcove, and I wondered: How many places on Earth still support the same roster of large mammals that were present there 500 years ago? Is Emas one of them? The question has an important relationship to rarity because 39 percent of large mammals with body mass greater than 20 kilograms (a maned wolf weighs in at around 23 kilograms) are considered threatened with extinction, as compared with 25 percent of mammals overall. Theory and lots of empirical data tell us that bigger mammals tend to be more wide-ranging than smaller mammals, and as George Powell had shown in Peru, most parks are too small to support them. With hunting of large mammals common almost everywhere today, the answer to our question, we speculated, could well be zero or at best very few places left on Earth with intact large-mammal faunas.

But the results of our research on the topic were more optimistic than we expected. The article we published in the *Journal of Mammalogy* in 2007 reported that 130 places on Earth serve as large-mammal refuges and that they support the full roster of big mammals that lived there 500 years ago. These refuges fell into two categories. One category included the most inhospitable places on Earth, places that were too cold, too damp, too dry, too humid, or too remote for humans to develop. This group featured vast regions of Siberia, northern Canada, Alaska, and the Amazon and Congo basins. The other group included a much smaller set of places, including Emas—remnants still afloat in human-dominated landscapes. We made no claims about whether the species survived at carrying capacity in these places—often they were present at much lower numbers or densities than in the past. It was clear, though, that places such as Emas would need intensive management to maintain the rich large-mammal fauna still present.

After lunch and a short drive, we arrived at a soy plantation that bordered the park. We stepped out of the car and into a completely different environment. Walking among the neat rows of soy, I asked Carly how many planters grew soybeans on farms adjacent to Emas. Gauging by my experience with rice cultivation in Nepal, I expected the answer to be in the hundreds. I was off by two orders of magnitude. Carly held up one finger. "This 40,000-hectare ranch is owned by one person. There probably aren't more than a few landowning entities in this entire area."

Big soy. It was my first encounter with such a vast expanse of agriculture. The lucrative plants covered the entire landscape. Corn grows here also, and in some parts of the Cerrado cotton is added to the rotation. The soybean is a recent addition to the farming economy of the region. *Glycine max*, as it is known to science, might have been more aptly named *Glycine "min"* until a few years ago, when crop and soil scientists figured out that the addition of lime to reduce soil acidity enabled the conversion of pasture and cattle ranching to soybean cultivation. As a consequence, the nitrogen-fixing legume began to prosper in areas where it could not grow before.

Brazilian farming practices changed almost overnight. Brazil has become the second-largest exporter of soy, after the United States. Much of it is exported as soybean oil, the most widely used cooking oil, or soybean meal, which has many uses but largely is fed to cattle. Producing high yields requires extensive use of fungicides, which are applied every few days. The corn grown in the next pasture was subjected to heavy doses of Roundup to control weeds. Pesticide use was rampant. White-tailed hawks and some kestrels flew by, but raptors and other birds were few. The agrotoxins may have already thinned their numbers. Perhaps a rumored decline in the bird fauna here, however anecdotal, is an early warning signal as to what lies ahead for surviving Cerrado wildlife.

That night at dinner with Leandro Silveira, Anah, and their team of researchers, the conversation focused on big mammals, beginning with rhinos and elephants in Asia and Africa. "Our biggest herbivore in Brazil is now the tapir," Leandro noted, "small by megaherbivore standards." A tapir is about the size of a large pony, with a pig-shaped body. Leandro was pointing out one of the great anomalies of nature. Even though there are more mammal species in South America than anywhere else, large mammals are even rarer here than in other parts of the world. In truth, most South American mammals can fit inside a shoe box. Of course, there are important exceptions that are considerably bigger—the larger primates, peccaries, capybaras and other large rodents, deer, and larger predators—all of which are well documented as having a major impact on their surroundings.

It was not always so. Paleontologists tell us that we are simply several million years too late to witness the South American Serengeti. Back in the Pliocene epoch, about 5.3 to 2.6 million years before the present, a rich megafauna filled South American forests, savannas, and pampas. Giant ground sloths rose up on their hind legs like giraffes to browse tree branches. Swamp mammals the size of hippos and rhinos crashed through the canebrakes. Across the grasslands galloped camel-like creatures. Around the waterholes lurked long-fanged marsupials that shared a common ancestor with opossums but bore a remarkable resemblance to saber-toothed cats.

The large mammals were basically done for before humans arrived, likely as a result of climate change. During most of the Age of Mammals—when the class Mammalia first evolved about 60 million years ago, after the ebb of the dinosaurs—South America was a continental island, and its unusual mammalian fauna evolved in isolation from the fauna of other continents. The first and probably greatest wave of extinctions occurred during the Great American Interchange about 4 million years ago, when the Isthmus of Panama rose above sea level, and resulted in a massively biased colonization. North American mammals flooded into South America and are now major components of the fauna (the maned wolf among them). Many fewer South American mammals became established in North America, and today only three species—the Virginia opossum, the porcupine, and the nine-banded armadillo—still survive there. Interestingly, two of the three species of our focus in this chapter—anteater and armadillo—are remnants of that native South American fauna.

When the renowned paleontologist George Gaylord Simpson came out with his 1980 classic *Splendid Isolation: The Curious History of South American Mammals*, he highlighted the divergent route South American mammals took in evolution because of physical barriers posed by the oceans and the flooded isthmus, as well as by the Andes, high deserts, and rain forests. These virtual fences prevented the mixing of the fauna. Thirty years after the publication of his work, intensive human activities including agricultural expansion, settlement, and transportation infrastructure, rather than mountain ranges or large rivers, have become the great isolators. The results, too, threaten millions of years of evolution. The IUCN Red List now includes many formerly common species that cannot persist in human-dominated landscapes, such as many antelope species and other large, hoofed mammals that used to migrate across large grasslands.

I asked Leandro and Carly to assess which of the mammals they were studying would survive in the face of expanding agriculture. I wondered if it was an ecological stretch for the grassland-loving mammals to occupy soybean fields, should landowners happen to allow it. Alternatively, I wondered whether some might be preadapted to colonize and persist in another open habitat, albeit one with a monoculture of soy, or on cattle ranches where the grass was cropped to the ground.

Carly's scat data showed that the answer varied, depending on the species. Maned wolves preferred Emas's grasslands and avoided closed-canopy forests inside and outside the park. Surprisingly, maned wolf scats were quite common out in the soy, however. She attributed the presence of the species there to the abundance of rodents, its dietary staple. Soy acted like natural vegetation and gave the rodents shelter. As long as they could go a-ratting, the wolves seemed to do fine in the croplands. Carly noted, "In spite of its shy nature, the maned wolf is adapting to expansion of agriculture." Maned wolves are common in Brazil's agriculture-dominated landscapes, at least those of soy, it seems, because they are not persecuted as is the gray wolf in northern climes. Maned wolves tend to avoid cattle ranchlands, though, as those lands are too bare even for rats.

In contrast, the giant armadillo had no use for soy plantations. Giant anteaters also avoided the agricultural fields, but they visited the cattle pastures because these appeared to have greater populations of ants and termites. Ranchers often left the termite mounds and ant nests intact, and the long-snouted opportunists showed no hesitance in excavating their dinner in man-made habitat. Giant anteaters were also more common where ranchlands abutted riparian forests. The shaggy creatures entered these forests and bathed in the water to cool down. Jaguars tended to stay inside the park but occasionally wandered into nearby fields. Pumas spent more time outside the park in areas with heavily forested groves lining rivers and streams. Tapirs liked to be near springs close to the remaining forest fragments outside the park.

Overall, the landscape matrix in this region was friendlier to rarity than most pessimistic biologists would have predicted. Some of the large mammals survived massive land-use change in the Cerrado because enough bits of natural habitat remained in the matrix to meet their needs for food and cover. How long these vital pieces would remain before becoming soy or sugarcane fields was an open question. Leandro was lobbying in favor of forever. He is one of the few muddy-boots biologists who has learned to rub elbows with lawmakers, and he spends as much time working with them as he does tracking jaguars. Leandro practices what the best conservation biologists preach, that conservation is 10 percent science and 90 percent negotiation. He had fought hard to enact the current federal law requiring landowners to keep a minimum of 20 to 30 percent of their property in natural vegetation, as well as to keep their hands off development of river- and streamside forests. Depending on topography and drainage systems, this measure potentially protects more than 35 percent of the Cerrado outside the reserves.

The Round Table on Responsible Soy Association, an effort to bring together all the big soy producers, conservation groups, and the Brazilian government, is trying to make cultivation more harmonious with nature conservation. Leandro's first step was to convince the group that the best-practices goal of leaving 20 to 30 percent of land as natural habitat would help the maned wolf persist in the Cerrado. It would be ideal if the protected areas were in contiguous blocks connecting ranchlands in one area to those in another. Carly worried that all of the set-asides would be in forests because including grassland in the 20 to 30 percent mix of intact land would bite into the profits of the ranchers. Under the current system, the big producers would benefit most by converting to agriculture as much grassland habitat as allowed. And that is what is happening. Even with a law that sounds good on paper, legislation that is not based on strong science will have uneven results. The current law could conserve good jaguar and tapir habitat outside reserves but do little for the grassland-dependent maned wolf.

The next morning, we were out early again with Mason. We stopped at a huge hole in the ground. Here were the signature diggings of a giant armadillo, one dedicated earthmover. "There must be an ant nest or termite mound nearby," Carly said. Giant armadillos often excavate burrows to reach under the nests of their favorite prey. Few people have ever seen a giant armadillo aboveground or in a zoo, so it's hard to picture one. A good start would be to multiply the size of the common nine-banded armadillo by ten. The nine-banded armadillo, the unofficial state mammal of Texas, is among the most common roadkill along highways in the state. If striking a poor nine-banded would feel like rolling over a speed bump, hitting a giant armadillo, which can weigh up to 60 kilograms, would be like smashing into a retaining wall. Fortunately, giant armadillos tend to stay clear of roads in the Cerrado, though some are run over by passing vehicles anyway. Unlike the nine-banded, which ranges across much of the southern and southeastern United States all the way to Argentina and parts of the Caribbean, the giant armadillo is limited to South America. There it ranges widely into the Amazon basin, where very little is known about its habits and needs. Most likely, though, giant armadillos, which are both rare and considered by the IUCN to be threatened, thrive in the drier portions of the range, in Venezuela, the Cerrado, and the Chaco region, which extends into Paraguay.

Carly and Mason found nearly sixty scats of this subterranean mammal. Together with the data from Leandro's team, which was the first to radio-track the species, giant armadillo ecology moved beyond the anecdotal phase. Even though tracked individuals proved to be largely nocturnal, this armadillo turned out to be the most sensitive of the large mammals to human disturbance. Not surprisingly, it shuns life in the soy fields. Leandro and his coworkers found that the home ranges of five individuals were about 10 square kilometers and put the density of armadillos in Emas proper at about 3.4 animals per 100 square kilometers, more akin to the densities of a top predator such as a jaguar than of a species that eats more abundant ants and termites. "Just like the anteaters, giant armadillos also have low birthrates," Carly mentioned. In the scientific literature, low reproductive rate is not itself viewed as an initial cause of rarity, but as we saw with the greater one-horned rhino, serious depletion of a population hampers rapid recovery and can leave the species more vulnerable to extinction.

At lunch, Carly and I discussed the peculiar habitats of the guild of ant- and termite-loving mammals. Mammals that subsist on the social insects are few in number and restricted to the tropics and subtropics, but their taxonomic spread is remarkable. Biologists call different groups that feed in a similar way an example of "convergence"; the various species that fill the termite-eating niche (or any analogous niche) are said to be "ecological equivalents." In South America, anteaters and armadillos, of the orders Pilosa and

Cingulata, respectively, are examples. Africa is home to the aardvark, the sole living member of the order Tubulidentata; the aardwolf, a hyena relative; and pangolins, or scaly anteaters, also in their own order, Pholidota. Asia also harbors pangolins and has the sloth bear, a true carnivore and a termite connoisseur. Australia and New Guinea have the spiny anteaters or echidnas, which represent the monotremes, and Australia's bandicoots represent the marsupials (though the most prevalent and species-rich termite-eating animals in Australia are not mammals but lizards).

Little is known about the ecology of pangolins, aardvarks, aardwolves, and sloth bears, but evidence indicates that where there are termites, the density of termite eaters can be impressive, even if their mostly nocturnal behavior makes them hard to see. Some, such as pangolins and sloth bears, are hunted for body parts used in folk medicine, and pangolins and armadillos are hunted for meat. Some of these species, including the sloth bear and some pangolins, are considered threatened by the IUCN. Without human pressures, then, the ant- and termite-eating contingent of the mammalian fauna might be even more common.

Back at the guesthouse, I had left my sneakers near where Mason was resting on the porch with his besotted curassow. "Be careful not to leave your shoes out at night," Carly warned. "The maned wolves steal them and chew them up for the salt."

I needed my shoes for the drive Carly was organizing for one of my last nights on the Cerrado, a trip to spotlight tapirs, crab-eating foxes, and the most important target, the maned wolf. Carly often searched along secondary roads, a mode of travel used by maned wolves that frequented the croplands. Giant armadillos, giant anteaters, tapirs, and pumas, however, avoided roads as much as possible.

The drive through the park grasslands turned up very few animals, but once we crossed over into searching the soy, sightings increased dramatically. Crab-eating foxes were everywhere and quite approachable. Nightjars, a kind of nighthawk also known as goatsuckers, sat on the roads that had been baked by the day's sun to stay warm in the chilly night. Carly said she had often spotlighted tapirs out in the Ag, but we saw none. More crab-eating foxes. These canids were certainly attractive, but we wanted the red fox on stilts, the maned wolf.

On the secondary road bordering the edge of Emas, Carly swung her spotlight in a wide arc. The driver slammed on the brakes, and all of us standing in the back of the open-air pickup lurched forward. A maned wolf stopped and then resumed sauntering along the road. For the next twenty minutes we motored along a road inside Emas, the wolf just outside on a parallel road bordering the soy. We were all surprised at how effortlessly the wolf kept up with the vehicle; we were moving at nearly 16 kilometers an hour, and the wolf seemed to maintain its graceful rapid walk. Finally, it slipped back into the soy. On the drive back to camp, the image of the dashing wolf stayed in my head, and it remains in my memory as vivid as ever.

⌒

Ten months later, in May 2010, I had a chance to stop in Seattle and again see Carly, who was in the last stages of completing her dissertation, and Mason, now officially retired (at the age of six) from his scat-searching expeditions. Carly had just answered one of her research questions with some final lab work on maned wolf hormones: Were the maned wolves that lived in the soy fields less healthy than those that stayed within the boundaries of Emas National Park? Carly bubbled with excitement over her findings. "The thyroid hormones from wolves in croplands are actually higher than from those in the park." High thyroid hormone levels were a good thing, and thyroid levels are useful indicators of nutritional status. The story emerging from Carly's work was that the maned wolves were selecting croplands for hunting rodents, partly because the hunting was easy and efficient. But there was more. Carly had discovered that the cropland wolves had higher levels of cortisol, indicating that the Ag land was also more stressful for that species to live in. So the soy fields may be an ecological trade-off for this elegant animal. Abundant food awaits in the soy, but so does the stress of leaving its natural habitat and venturing into the human landscape, where perhaps it feels more exposed and vulnerable.

We compared notes about how natural landscapes were being transformed in the places we had been working. I mentioned the vast plantations of oil palm and acacia (cultivated for pulp) I had just seen on Borneo and Sumatra. "Remember the soy plantation we walked through looking for wolf scat?" Carly interjected. "In one year, it changed over to sugarcane."

Does it matter to rare grassland mammals if the agricultural expansion involves sugarcane, soy, tapioca, or cacao? To the Cerrado trio, the answer is yes. The conversion of low-growing soy to sugarcane will shake things up ecologically, Carly explained. With the soy, there is a relatively welcoming landscape matrix for the wild animals in the region. "But when converted to sugarcane, it becomes inhospitable to an open vegetation–adapted species like my wolves." Compared with soy, sugarcane is a longer-growing crop between harvest periods, and it forms tall, dark stands, which entails loss of the open-landscape qualities mimicked by soy fields that are conducive to species such as the maned wolf. Sugarcane also requires more manual cultivation, and factories may be built to handle the cane on-site, so it could become a magnet to draw in workers. Carly didn't sound hopeful. "I see conflict as animals try to avoid humans, and as poachers seek out the wildlife that wanders into the Ag."

In Sumatra, I had spoken to David McLaughlin, an agricultural expert for the World Wildlife Fund who had worked in tropical agriculture all over the world. I asked him how agribusiness can maintain, year after year, the commodities we use without tropical nature evolving innovative pathogens and leaf pests to attack the widespread cultivated plants. Every crop plant has legions of pests, some minor, some fatal if not controlled, he replied. "For your oil palm, it's lethal spear rot, found in South America in the 1980s but not yet present in Southeast Asia. Cacao trees have all sorts of things that can hit them. And tropical soy needs constant vigilance against fungi and herbivores."

"So," I continued, "you mean if some careless biologist introduced a pathogen into a soy or oil palm plantation, it could wipe out the whole place and native vegetation would return?" David eyed me suspiciously. What Carly had observed around Emas foreshadowed his answer. "Remember, commodity production can turn on a dime. Even if a crop pest attacked soy, farmers could shift immediately to growing sugarcane for biofuels. And if the sugarcane succumbed, they might go back to cattle ranching or corn." The point is that land once converted to agriculture will likely never go back to native vegetation but rather will be converted to the next big monoculture. The same holds true in the vast oil palm developments in Southeast Asia. Those that fall to some new pathogen will probably be replaced by acacia trees for pulpwood or return to rubber. The commodities are never stable, but the pressure applied by the growing human footprint is mounting. In the meantime, programs designed to enhance wildlife-friendly certification of major cash crops—whereby consumers know that production of the crop has avoided the conversion of natural habitat to produce it—is still years away.

The Cerrado story helps us put countryside biogeography, or matrix conservation—or whatever term we create to describe the landscape approach to conservation that considers species both in parks and in the human-dominated areas outside of formal protection—into a larger context. Many species that do fine in wild habitats can persist in adjacent farmland dotted with pockets of natural vegetation. This is especially the case where riparian zones are protected from development. But for how long? Perhaps our assumption that various species will persist in agricultural lands is an ecological mirage: it may be that they will merely hang on for another decade or two and then start to decline as a result of pesticide residues, stress, or other sources of mortality. Furthermore, species that survive in human-dominated landscapes tend to be generalists, not the specialists or the rarities that this book portrays. Carly believes that upholding the Forest Code of Brazil is key, that it is these scattered remnant habitats that enable species to use the landscape as a whole. The future challenge of agroecology is to identify the opportunities for mutual accommodation and its limits—in the Cerrado and elsewhere as well.

The scale at which the face of Earth is being converted from natural habitats to cultivation and the pace of it are truly staggering. Global projections are that with 9 billion people on board Earth, world food demand will double by 2050, and the Cerrado will play a key role. Yet most people are unaware of the enormity of agriculturalization and other habitat encroachment. Various small-scale local accommodations, even if increased in number, may not be enough to protect wild nature. A hidden aspect of this problem is the scale of buy-ups of land in the tropics and elsewhere by corporations and nations as they position themselves for the future of their food supplies.

So the problem grows larger: wholesale conversion of land not only threatens to make no small number of common species rare through human activity; it also threatens the very existence of what is now rare. One has to hope that the rare species of the Cerrado and other areas of intensive cultivation are more adaptable than we think and that efforts to enable coexistence on working lands, by habitat protection and then by best practices on and near agricultural lands, will enable persistence. It would be a shame for others to lose the chance to observe the remarkable silhouette of a giant anteater or the outline of a graceful maned wolf on a moonlit night in the Cerrado.

Chapter 7

Invasion and Resistance

*A*BOUT A CENTURY AGO, SUGARCANE PLANTERS in Hawaii faced major crop losses from a teeming rat population. Their solution seemed practical at the time: import the South Asian common mongoose—by way of Jamaica—to eradicate the vermin. The planters unfortunately selected the wrong control agent, with disastrous consequences. The diurnal mongooses were exposed as abysmal predators of the nocturnal rats (which at an earlier time had also been introduced into Hawaii) but avid consumers of the eggs and nestlings of native birds, many of them rare and found nowhere else in the world. It was too late to correct the mistake; the mongooses had been set free on all the main islands except Kauai. En route to that island, a wise official is said to have dumped the mongooses overboard. Stuart Pimm, an expert on Hawaiian extinctions, corrects the account in this way: "The mongooses intended for Kauai bit the fingers of the boatman and he drowned them. That's why they never made it there."

Hawaii's imported rats and mongooses are part of a group that ecologists term invasive exotic species. These interlopers have the ability to outcompete native flora or fauna when they are introduced into an ecosystem where they do not occur naturally. Every continent has them; in North America, some of the worst invasives are plants such as purple loosestrife from Europe and kudzu from China. In Nepal, the explosion of mile-a-minute vine from South America threatens to drape over much of the greater one-horned rhinoceros's prime feeding areas. Rabbits and red foxes are invasive exotic mammal species in Australia. Some species come to or are brought to a land and have no effect on the local flora and fauna, but others crowd out or kill off native species, not infrequently rare ones, and those obviously are the ones of great concern. Introductions of mammals onto islands—giant ones such as Australia or small archipelagoes such as Hawaii—have nearly always been disastrous because mammals, as we saw in Peru and Nepal (chapters 3 and 5), are the prime ecosystem engineers everywhere.

Fortunately, most mainland mammals—be they mongooses or monkeys—and many reptiles and amphibians and even ants rarely occupied distant islands before human navigation. The exceptions are continental islands that were once connected to a large landmass, such as those islands that were part of the Sunda Shelf, once connecting Southeast Asia that sits west of Wallace's Line, and another once linking Australia to New Guinea by a land bridge (see chapter 2). There are good reasons why this is so: most terrestrial creatures simply can't swim far enough or survive the passage from the mainland by clinging to a raft of floating vegetation, although snakes and lizards are much better at it than mammals. Terrestrial mammals (with the exception of bats) and amphibians are especially poor dispersers over marine barriers for another vital reason: like humans, if they drink seawater they lose more water in trying to excrete the excess salt ingested. Here, two lines from Samuel Taylor

Coleridge's *Rime of the Ancient Mariner* are as apt as basic physiology to explain the conundrum: "Water, water, everywhere, / Nor any drop to drink." In the absence of most terrestrial mammals, different ecological worlds evolve. For example, there are places without large cats as top predators, such as on the islands of Komodo and New Caledonia. Komodo has a dinosaur ecosystem where a large monitor lizard has become the apex predator, even hunting mammals. In New Caledonia, in the absence of mammalian predators, the geckos of the island have undergone a divergence of species, the largest of which has become the "tiger" of the island, eating smaller vertebrates.

Even the most remote archipelagoes that, like Hawaii, have never been connected to a mainland, however, experience waves of potential invasions by various species. Some come as stowaways on human-powered vessels, like Polynesian rats, or, like the mongoose, are deliberately introduced. But dispersal is a natural process that happens without human assistance. Some colonists do survive on floating mats of vegetation, carried by the ocean's currents to their new destinations. The vast majority of natural colonizations fail, however. Most species drown en route or die of dehydration, as already noted, or starve to death soon after arrival, or fail to find a mate or to leave enough offspring to carry on. This chapter focuses on the devastating effects of successful invasive species and the ecological disruption they cause to island life, especially the rarities endemic to the islands.

Invasive species can disrupt local ecologies anywhere, but for several evolutionary reasons they often hit remote oceanic islands especially hard. Being a great disperser allows a species to colonize a remote island such as Hawaii. But once there, in the absence of significant predators, natural selection strongly favors against dispersing offspring because they end up drowning if they head out to find new places to live and breed. One outcome is that on oceanic islands flightless birds and insects have repeatedly evolved. A paradox of nature: once arrived on such an island, it is better to become sedentary, but as a result, those species that do are more vulnerable to predators that are later introduced.

The flora and fauna on islands recently connected to the mainland—as with Sri Lanka to India or Trinidad to South America—are similar to those on the nearby mainland, so newly arriving species must cope with predators and herbivores they encountered on the mainland. But species that established themselves on islands that were never connected to the mainland or were otherwise far from continental coasts typically evolved in the absence of predators. The "naive" resident fauna lacked appropriate escape behavior, even birds that could simply have picked up and flown away at the first hint of danger when predators, including *Homo sapiens*, were introduced. As a result, most vertebrate extinctions during the past several centuries have occurred on islands. Such ecological naïveté extends to plants, too. As we shall see in the case of Hawaii, plants that evolved over time in the absence of large herbivores lost their chemical defenses and physical structures such as thorns and prickles; they suffer terribly when exposed to introduced goats, deer, sheep, and rabbits. An often overlooked group of exotic, invasive organisms are invisible to the naked eye—bacteria, viruses, and protozoans that cause outbreaks of disease. In some areas, introduction of disease-causing organisms has been even more deadly for rare natives that have had no previous cause to develop antibodies than have been creatures with backbones. Hawaii has been hard hit on the microscopic front as well.

So much attention has been given to recent invasions as part of the contemporary biodiversity crisis that we lose sight of the fact that some ecologically catastrophic invasions happened long ago. The first biodiversity crisis in Hawaiian history occurred when the Polynesians arrived more than 1,000 years ago, bringing with them slash-and-burn agriculture, pigs, and stowaway rats that scurried to shore. Combined with the unwelcome ark of more recent invaders that Europeans introduced—sheep, goats, cattle, deer, mongooses, and mosquitoes—these foreigners now plague an archipelago with perhaps the highest concentration of rarities on Earth. Ironically, while the suppression of malaria played a significant role in the decline of the rhinoceros in Nepal, the introduction of malaria in Hawaii has decimated a group of native birds, the honeycreepers.

That some of the spectacular native species are still holding on, such as the woodpecker-like *'akiapōlā'au*, demands our attention. Many evolutionary biologists, such as Stanford University's Peter Vitousek, a native son, view the string of South Pacific islands collectively called Hawaii as America's Galápagos Islands. If Hawaii's multitude of endemic species has often been decimated by invasives, this cradle of evolutionary exuberance, like Darwin's living laboratory, is also a showcase for how natural selection and isolation have combined to foster new species, many of them rare. Hawaii supports the highest percentage of plants and animals found nowhere else. For example, of the nearly 1,000 species of flowering plants found in Hawaii, 90 percent are restricted to the islands. This unusual level of endemism and such narrow ranges means that if we lose a piece of wild Hawaii, we lose many species forever.

In chapter 2, we looked briefly at the process of adaptive radiation as it occurred among birds of paradise in New Guinea. Evolutionary biologists have also discovered in Hawaii an extraordinary array of species created through adaptive radiation, such as honeycreepers, ferns, tarweeds, and the two most prolific but least heralded radiations, that of the land snails and the fruit flies. At least fifty-one species of Hawaiian honeycreepers, the main protagonists in this chapter, evolved from a single species of primeval finch originating on the American mainland. Eventually some honeycreepers, such as the *'i'iwi*, evolved a long, decurved bill to feed on nectar from species such as *Lobelia* that have appropriately curved flowers. Other finches evolved a facility for eating fruit or for gleaning insects from bark. Still others stayed with the ancestral finch behavior of cracking seeds. Some of these evolutionary variations never spread beyond a single island. Members of some bird species, though, managed to make it to another island. The *'i'iwi*, for example, populated several islands. Many of these species are now quite rare, and at least twenty are extinct.

E. O. Wilson has shown how the arrival of a founding species—such as the prototype of the honeycreeper—is typically the opening to an evolutionary play told in four acts. In act 1, a species arrives and colonizes the entire archipelago, expanding in range and numbers. By act 2, populations on the different islands have become distinct from the founding father and mother, and on some islands the populations become extinct. By act 3, most of the original birds to settle on various islands in the archipelago have gone extinct, and those that remain have evolved into distinct species, as different as the 'akiapōlā'au is from the 'i'iwi. Finally, in act 4, only a single population remains as an endemic species limited to a single island. The legacy of a single, widely dispersing species colonizing many islands in an archipelago is the generation of many new species with very narrow ranges.

This process of adaptive radiation is common on island archipelagoes because radiation from a single ancestor and formation of new species require barriers to gene flow between forming species, often provided by water gaps among islands. There probably would have been no radiation of finches in either Hawaii or the Galápagos if there had been only one island of a size equal to the sum of current islands. Such avian speciation did happen in New Guinea, a single island, only because New Guinea is so large and the steep mountain chains are sufficiently isolated from one another that they served as barriers to commingling, just as the currents and deep water separating archipelagic islands often do.

When it comes to organisms with more limited mobility and range, it's a different story. Even the main island of Hawaii is large enough to witness the explosive radiation of two diminutive and relatively weak dispersers, fruit flies and snails. Speciation in birds, and even in land snails and fruit flies, requires geographic isolation of populations for long enough that they accumulate sufficient genetic differences that they fail to interbreed if and when they subsequently come into contact. One size does not fit all, however. What is a dispersal barrier to a snail or a fruit fly is not a barrier to a bird.

On isolated island chains such as Hawaii, then, one can witness both the results of incredible Galápagos-like radiations from a single ancestral species and the catastrophic effects of some more recent introductions by humankind. The outcome of this clash, the first evolved over eons, the other more recent, and its effect on rarity has application to oceanic islands everywhere.

~

I had a chance to see honeycreepers during a visit to Hawaii's Big Island in July 2001. Ecologist David Wilcove, a friend who had also accompanied us on part of the journey in the Cerrado, was with me. Our main purpose was to attend an international conference on conservation biology taking place at the University of Hawaii's Hilo campus. The real attraction, however, was a chance to risk burning our retinas by looking straight at a molten-red 'i'iwi, perhaps the most beautiful of all Hawaiian honeycreepers, and to meet scientists in charge of saving the extant half of this endangered family.

At dawn on the first day, David and I headed up the road to the pass between Hawaii's reigning natural monarchs—the volcanoes Mauna Kea and Mauna Loa. Joining us was conservation planner John Morrison, a colleague from the World Wildlife Fund; later that afternoon we would rendezvous with another, ornithologist John Lamoreux. We ignored the danger posed by the drivers hurtling by at high speeds and instead took pleasure in listening to one another mangle pronunciation of the local bird names. If we could spot even one of the birds whose names we were trying to master, such as the 'i'iwi (pronounced e-*e*-vee), we joked, our struggles with the Hawaiian language would disappear. We stopped at a popular birding spot, K puka 21—a *k puka* being a forest patch that has become isolated by a recent lava flow. Native koa and *'ōhi'a* trees make up these fragments of mature forest, the 'ōhi'a sporting flowers

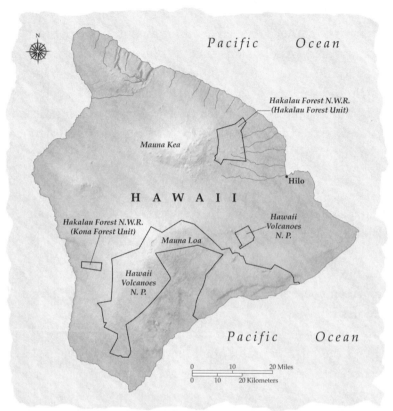

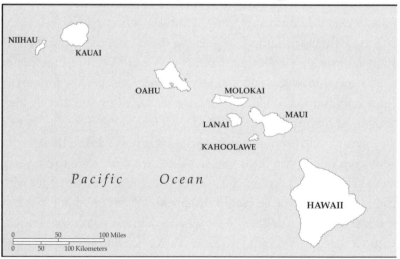

Map of Big Island (Hawaii) and its position within the Hawaiian archipelago

resembling the bottlebrush blossoms of eucalyptus. The bright red color of the flowers acted as a magnet for the nectar-feeding ʻiʻiwi and ʻapapane.

We picked our way over treacherous crumbles of hardened lava to reach the forest edge. Within moments we saw what at first glance looked like a vermilion-colored ʻōhiʻa flower. Suddenly the flower started to move. The dense flowers and foliage only partially concealed the scarlet and black plumage of the gorgeous bird that, once fully exposed, was ignited by the bright mountain sunlight like a flare—an ʻiʻiwi. Iʻiwis also occur in native forests on Maui, Kauai, and Molokai, but they are now almost extinct on Oahu; today fewer than five are left. A visitor could spend a day, as I once did in the early 1990s, driving around the lowlands of Oahu, un-successfully trying to spot a single honeycreeper or native plant. Had I read more about Hawaiian ecology in advance, I would have saved some gasoline. Many native Hawaiian birds and plants that are not yet extinct are fugitives, finding refuge on islands such as Oahu only in mountain strongholds. On the Big Island, the birds are more accessible.

We continued in search of the palila, another honeycreeper spe-cies that resembles in its behavior and profile the pine grosbeak of North America. Having emerged from the rain forest, we were now about 66 kilometers from Hilo, in the Big Island's dry forest zone, or what is left of it. Cattle ranches lined the road; such pastures are widespread in tropical dry forest areas globally. The climate and soils are more conducive to raising livestock than are the pastures carved out of true rain forest and abandoned a few years later. The spread of cattle on the Big Island has turned a continuous belt of dry forest composed of native *māmane*, a leguminous tree, and naio, a kind of sandalwood tree, into a savanna with an understory of in-troduced grasses. Today only about 1,200 palilas remain in the wild, and the species has become the subject of a US Fish and Wildlife Service recovery program. We spent the next four hours walking along the road on the slopes of Mauna Kea where, according to our birding guide, the palila was supposed to be, searching māmane branches for a sturdy yellow-headed honeycreeper. No luck.

The endangered palila population is an example of what can happen when a rare species becomes too dependent on a particular food plant. The māmane is manna for the palila, and when the trees disappear—as a result of pigs browsing the seedlings or the woodlands being cleared for cattle ranches—the birds vanish. The fate of the māmane is a prime example of the potential ecological repercussions of introducing large plant-eating mammals, such as cattle or pigs, onto an island where they never occurred naturally. The local species of plants were unprepared for the invasion and had not evolved chemical defenses to deter herbivory. There are also larger evolutionary processes at work here: the birds are so specialized on māmane in part because the Hawaiian flora has become impoverished and in part because the palila does not venture beyond a narrow range at midelevation on the Big Island.

Fortunately, on our way back to Hilo we took a wrong turn. Quite by accident, we found ourselves facing a feeding party of palilas and an 'elepaio, the endemic Hawaiian descendant of an Australasian flycatcher. As we enjoyed the spectacle we had hoped to see, we also scanned the panorama of hills and fragmented forest perched above the converted, sugarcane- and pineapple-planted lowlands. The palila's strategy for survival became obvious, mirroring what all resistance fighters know instinctively: when the invader advances, head for the high ground—in this case, where the māmane forest was still present.

To improve our chances of seeing native Hawaiian fauna and flora, we continued a search of Hawaii's uplands, choosing Hakalau Forest National Wildlife Refuge. Our group was fortunate to have as a guide Jack Jeffrey of the US Fish and Wildlife Service, who has been the force behind conservation at Hakalau for several decades. A young scientist studying the birds of Hakalau Forest, Liba Pejchar, now an assistant professor at Colorado State University, also served as a guide. Liba's specialty was none other than the 'akiapōlā'au—'aki for short—the renegade honeycreeper that had evolved to become the island's de facto woodpecker. For several weeks, she had been watching a male make feeding trips to a female and their young. Although it feeds like woodpeckers and sapsuckers, the 'aki does not nest in tree cavities; instead, it makes a cup nest in the terminal branches of the 'ōhi'a tree. If we were lucky, our group would see this rare honeycreeper, but to find even one of the remaining 600 pairs spread over several thousand square kilometers in only three hours would indeed require a blessing from the birding gods.

We walked through a stand of tall 'ōhi'a and giant koa trees—Hawaiian old growth—which gave way to meadows and back again to forest. For a biologist and lover of fantasy fiction, our hobbit-like experience of walking through open glades interspersed with groves of giants with their odd architecture was like a walk through Tolkien's Middle Earth. It's an ancient landscape, too: some of the 'ōhi'as have been carbon-dated at over 400 years of age, making them possibly the oldest nonconiferous trees in the entire United States.

Jack's knowledge of Hawaiian natural history flowed nonstop. He pointed out that the lack of native mammalian herbivores had allowed native vegetation over evolutionary time to lose defenses against being eaten: spines and thorns and waxy leaf cuticles and other chemicals that make plants taste bad. Thus, the native raspberry in Hawaii has many fewer and softer thorns than do raspberries elsewhere, the native nettle has lost its stinging hairs, and the local mints no longer produce the strong aromatic oils that emit their characteristic fragrance in other lands. David crushed the leaves of a nearby 'ōhi'a, expecting the small, fuzzy foliage to release a strong guava-like aroma. This dominant Hawaiian tree is a member of the eucalyptus family, known for the powerfully volatile oils infusing the trees' leaves and bark. He inhaled deeply, but there was no smell. Like other native Hawaiian plants, the 'ōhi'a no longer needed to repel large browsing mammals as had its evolutionary ancestors, and so, over time, its costly perfume factory had shut down.

Koa (*Acacia koa*), the other old-growth tree species here, is one of a strange-looking group of acacias. Most wildlife tourists associate "acacia" with the umbrella thorn acacia, *Acacia tortilis*, the flat-topped tree of postcard East Africa. Most amateur botanists recognize the genus *Acacia* by its bipinnate compound leaves and wicked thorns. These features aren't present in the koa, though. That's because there aren't any leaves, at least not on adult koas. Instead, koas photosynthesize with structures botanists call phyllodes—flattened petioles, or leaf stems, that serve the purpose of leaves and resemble small green boomerangs. *Acacia* species with phyllodes instead of leaves are a phenomenon seen only in Australia and the Pacific islands. Nearly all Australian acacias lack thorns and photosynthesize with phyllodes. And with no giraffes or elephants in the Hawaiian fauna, koa trees, of course, didn't need thorns either.

We started to see clusters of 'i'iwis and a number of 'apapanes, the other red nectar-feeding honeycreeper. Along the way, Jack pointed out a tall, familiar-looking plant with large greenish-purple curved flowers. It was a lobelia—related to the cardinal flowers American birders grow to attract hummingbirds. This was no common cardinal flower, though, but one of the rarest plants on Earth. *Clermontia pyrularia* had nearly gone extinct, but Jack and his colleagues had managed to rescue it. Now this species and other lobeliads are propagated and transplanted back into the wild. Jack and his colleagues are focusing the same kind of intensive restoration efforts on some rare mints that Jack discovered. Some of the other honeycreeper-pollinated plants, like the lobelias, are now imperiled because they have lost their pollinators and must be hand-pollinated in the field to keep the populations alive.

When I asked Jack about restoration of Hawaiian silverswords, his eyes lit up. Silverswords, which take their name from the long, narrow leaves and silvery hairs found on most species, are a

Hawaiian rarity that some plant conservationists have devoted their lives to saving. These plants can live to be fifty years of age, show remarkable adaptations to cope with rarity, and have an odd life cycle, waiting until the end of their life to send up a flowering stalk and then fruit before dying. Silverswords, which also go by the far less poetic name of tarweeds, belong to the genus *Argyroxiphium*, a small group of five species in the sunflower family. But like the honeycreepers, silverswords radiated into an array of more than thirty species. This group is known as the silversword alliance, a brother- or sisterhood of plants adapted to living under extremely harsh conditions—under the volcano and on its cinder fields or in acid swamps known as bogs. How could plants that survive the tough life in the shadows of volcanoes, exposed to wind and freezing temperatures and desiccation from intense sunshine, and those that thrive in bogs and in such low-nutrient soils, have become so rare? Volcanoes and bogs should be the boot camps for rare plants, toughening them up and allowing them to resist any disaster, manmade or natural. Silverswords even show peculiar adaptations enabling them to raise their body temperature by focusing sunlight on the shoot tips. What they are not adapted to is the trampling of pigs and browsing of goats that have been introduced here. Their thinly buried roots, especially those of the bog-loving species, are easily destroyed, and their succulent leaves are prime delicacies for goats.

Even before humans arrived on the islands, all species of silverswords had very limited ranges, and they remain vulnerable to extinction. Some populations are quite numerous, such as the Haleakala silversword, found only in an old crater on Maui. Its entire global distribution is limited to this one site, but this tiny area supports more than 40,000 individuals. Yet however common this species of silversword may be locally, all its eggs, or seeds, are in this one basket. Lose this site to pigs and goats and you lose the species forever. Jack and his colleagues, in addition to being botanists, have become world-class fence builders and goat rustlers, courses never taught in the ivory tower but essential to study and save rarities in the wild.

Scanning the trees, Jack and Liba spotted a honeycreeper we hadn't yet seen that day. The *ʻākepa* dazzles beholders with its burnt-orange plumage and slightly crossed bill, which enables it to pry open leaf buds for prey. Suddenly the sounds of the native birds were drowned out by the exotic red-billed leiothrix and a flock of Japanese white-eyes. One person asked, "If you erected a mist net to capture birds, would you catch more native or exotic species?" The ratio, Jack replied, would be about ten individuals of native bird species for every introduced one. We were relieved to hear that, in an archipelago full of things from everywhere else, native Hawaiians still reign in Hakalau Forest.

When our party stopped for lunch, a native fruit fly landed on a bush near David, perhaps attracted by his fruit salad. In Hawaii, fruit flies form new species like an out-of-control experiment. Since any individual fly could be new to science, it was tempting to thrust it into alcohol and send it on to a fly taxonomist. Identified so far are five hundred species of fruit flies that swarm wildflowers all over Hawaii, but some experts estimate that the number of known species will increase. Fruit flies are not particularly endangered, as are many honeycreepers, but like them some are quite beautiful and ornate, far more compelling than the red-eyed, vestigial-winged mutants of genetics class. Other insects were about, but if we looked closer, we would find only introduced ant species grubbing for our lunch spoils. Ants never made it to Hawaii before human introduction, and many of the arthropods that did arrive, like many birds, shed their wings to become flightless versions of their mainland ancestors. Hawaii has a depauperate collection of insects in terms of number, but those that do occur on the islands are highly endemic and often have narrow ranges. Remarkably, the Hawaiian Islands have more species of fruit flies than are found in the rest of the world combined. "First in Fruit Flies"—not a slogan that will appear anytime soon on automobile license plates in Hawaii, but of profound interest to biologists.

First honeycreepers, then silverswords, and now fruit flies are showing us how rare species persist in a compressed space and how adaptive radiation results in many new species that themselves are rare. Adaptive radiations have occurred in many places—in South America and Australia and even in Mexico's Chihuahuan Desert, home to a great radiation of prickly pear cacti, evening primroses, and desert fish (cichlids), wherever the necessary conditions have prevailed. But it is easier to study radiations on island archipelagoes. The Galápagos Islands are home to radiations of finches, mockingbirds, tortoises, and giant dandelions.

The fruit fly kept hovering about; evolutionary marvel or not, I would have traded encountering ten new species of fruit fly for a single 'akiapōlā'au. We resumed our walk, hoping to catch sight of the faux woodpecker. Liba showed us the nest, to which the male typically returned about every forty-five minutes to feed the female. Our plan was to sit silently under the tree and wait. A half hour into the vigil, Liba stood up and pointed to her left. She had heard an 'aki calling in the distance. We heard nothing except red-billed leiothrixes, house finches, and white-eyes, all imported singers.

The hour was late, so we had to end the stakeout. Jack suggested that everyone come back to Hakalau Forest during the breeding season (December through May), when the native birds would be more active and vocal. Then the trees would be "dripping with 'i'iwis," and maybe an 'aki or two. Liba chimed in: "People expect all honeycreepers to be rare everywhere, given all the threats they face. But the dawn chorus at places like Hakalau Forest is unbelievable; it's deafening. I think songbirds are at higher densities here than almost any forest I've visited." Honeycreepers have declined in the lowlands of all of Hawaii's archipelago partly because the dense forests of native plants have been replaced by a generic pantropical vegetation including eucalypts from Australia, banyan trees from India, and jacarandas from Brazil. On Oahu, if you relied on the most common plants to distinguish where you were in the world, you might think you were in Bermuda, or St. Croix, or even a Club Med in the Seychelles. Most of the trees and shrubs that line the roadside display large, showy flowers in an orgy of colors, imported beauty that has increased the diversity of plants in Hawaii but fails to attract its native birds.

Introduced species of birds and mammals are also responsible for pushing some of the rare natives out of the lowlands. According to Thane Pratt, the leading authority on Hawaiian birds, more than 170 alien bird species have been introduced onto the Hawaiian Islands, about 54 of which have established breeding populations. Like their floral counterparts, they came from the world over—red-billed leiothrixes, rose-ringed parakeets, and bulbuls from Asia; waxbills and canaries from Africa; cardinals and meadowlarks from the Americas. These and most of the other introduced avifauna are all beautiful birds, but they don't belong in Hawaii.

The most pervasive factor accounting for the absence of native birds below 1,500 meters' elevation is unexpectedly small. The accidental introduction of mosquitoes in the early 1800s allowed avian pox and avian malaria to spread from infected migratory birds or species introduced from Asia to the defenseless native birds, wiping out a number of species and populations in the lowlands. The remaining species were those that ranged into the higher elevations, where it is too cool for the *Culex* mosquitoes to survive.

To imagine the 'aki and the other honeycreepers joining the dodo, the great auk, and the elephant bird in the aviary of oblivion is a painful thought. Jack is concerned that with rising temperatures resulting from global climate change, the temperature-sensitive malarial parasites could develop fully within their mosquito hosts. Then the disease could make its way up the mountainsides and adapt to life in these forests, wiping out the remaining native birds as it goes. (Studies of human malaria show its projected spread to higher elevations and latitudes for the same reason.) Birds at still higher elevations might be safe for a while, but how much farther uphill can they move before the forest gives way to the cinder fields characteristic of the volcanic peaks?

The observatory by the summit of nearby Mauna Loa gives us some context for contemplating that question. The Mauna Loa atmospheric carbon dioxide (CO_2) measurements, which began in 1958, constitute the longest continuous record of changing atmospheric CO_2 concentrations. Earth's atmosphere mixes globally so that the CO_2 component at Mauna Loa, where there are few local inputs to the atmosphere, is a good gauge of the overall fraction. The methods and equipment used to obtain these measurements have remained essentially unchanged during the five-decade monitoring program. The Mauna Loa record shows a major increase in the mean annual concentration of CO_2, from 316 parts per million by volume (ppmv) of dry air in 1959 to 385 ppmv in 2010. In short, the increasing amounts of CO_2 in Earth's atmosphere are real and serious.

Many scientists have warned about the spread of tropical diseases with shifts in climate, but few have considered its impact on particular species of wildlife, especially rarities. This is an issue that extends far beyond Hawaii. Species are beginning to head uphill to find cooler and often wetter refuges, and as the climate warms further, their chances of success are likely to decline. Shifting one's range up a mountain is especially problematic for organisms other than birds because of limited mobility options.

~

We returned to Hilo in time for the opening plenary lecture by David Steadman, an authority on extinctions in the Pacific region. Part-time comedian and full-time paleontologist, anthropologist, and evolutionary biologist, he related an ecological history that captivated many in the audience who were new to the region. Steadman projected a map of the Pacific Ocean on the screen. Pointing out the island groups in the order that they were discovered by Polynesian sailors, he showed that the Hawaiian Islands were among the last. Anthropologists generally date the first landing on Hawaii by the Polynesians (who hailed from the Marquesas Islands) around AD 1200–1300, although some experts set their arrival more than 800 years earlier. The prevailing winds, which skirt the Hawaiian archipelago, had discouraged human settlement longer than had been the case anywhere else except New Zealand and Easter Island. After the initial discoveries, several more migrations to the Hawaiian Islands occurred from Tahiti and elsewhere. Along with them came the normal passengers for colonization and exploration—taro root, bananas, coconut palms, dogs, hogs, and chickens—and the stowaways, the Polynesian rats.

When Captain James Cook arrived in 1778, the total human population was less than 1 million. But by then the biological effects of the Polynesian invasion were well established in Hawaii, just as they were wherever else the Polynesians had landed across the Pacific.

Polynesian farming practices involved widespread clearing and burning of the lowlands, a process that wiped out those species adapted to life along the coast. New information suggests, however, that the Polynesian rat may have played a bigger role in the demise of the native birds than did land clearing. The naïveté of native birds that had evolved in the absence of predators such as rats had left them defenseless. The birds weren't able to migrate or to evolve quickly enough ways to protect their nests. So the rats ate the eggs and nestlings, reproduction plummeted, and many species passed into oblivion.

Steadman's current focus is across the South Pacific, where he spends his days in bird graveyards reassembling the rich faunas that were clubbed to death by hungry Polynesians and exploring lava-isolated pockets where the fossils are the result of natural mortality. His observations illustrated the importance of an ecohistorical approach to understanding contemporary rarity and extinction. For example, his research sheds light on the extent and timing of the start of the biodiversity crisis on Pacific islands by showing us that a steep decline in species occurred nearly 1,000 years ago. It also offers clear evidence that natural rarity was exacerbated by invasion of humans and the fauna they brought with them.

After Steadman's talk, I asked Thane Pratt about the list of obituaries and the percentage of those species that were flightless. He listed 5 species of geese (plus 1 not yet described by science), 2 ibises, 1 hawk (possibly 1 undescribed), 1 eagle, 5 rails (plus around 5 undescribed), 4 owls, 2 crows (plus 2 undescribed), 18 honeycreepers, and more recently a flightless duck from Kauai that went extinct, for a total of nearly 40 extinctions. Of these described extinct species, 13 were flightless, to which could be added 6 undescribed flightless species, for a total of 19. Add the historical Hawaiian and Laysan rails and you have 21 species that had lost their ability to fly to safer ground when faced with a new threat and that had perished as a result. That's only a short list of what was lost before the arrival of Europeans, based on the species paleontologists have identified from several excavations scattered among the islands. The true number of extinctions remains a guess.

$\sim$

Sitting in a darkened auditorium listening to one scientific talk after another can induce slumber. So much talk about extinction also rekindled my desire to get out and see these rarities before they disappeared. Later that morning, our bird-seeking group returned to the upland forest along the Saddle Road, where Liba once again was our guide. A tip steered us to the Pu'u 'Ō'ō Trail for spotting 'akis. "*Pu'u*" means hill, and "*'ō'ō*" is the name of an extinct Hawaiian bird that represented an endemic family, closest to the silky flycatchers and waxwings. The 'ō'ō must have been a dashing bird, a jet-black honeyeater more than thirty centimeters long with yellow epaulets and undertail coverts. So coveted was its plumage that many were slaughtered for the feather trade, reportedly as many as 1,000 individuals around 1898 just outside of Hilo. By 1987 the 'ō'ō had been wiped out, the victim of habitat loss, hunting, and disease, illustrating once again that there are usually multiple causes, not just one, for such an ultimate fate.

A kilometer along the trail in the upland forest, we passed through groves of koa. *'Amakihis* and 'apapanes were about, but no 'akis. Then suddenly Liba signaled that she'd heard the song of the 'aki. We bushwhacked for several hundred meters to where the 'akis were calling, and there the birds sat, perched in a koa. An 'akiapōlā'au could appear as easily in an illustrated book of fairy tales as in a field guide. The color of the male's plumage adds a brilliant new interpretation to the color yellow. The 'aki uses its straight lower bill to hammer at tree bark like a woodpecker. Then it uses its long, thin decurved upper bill like a one-armed tweezer to pull out tasty grubs. David Wilcove describes the 'aki as a bird with "a Swiss army knife for a bill."

"That's not the only foraging trick up their sleeves," Liba explained. "They can also thrust out their lower bill so that the tips meet and use their mandibles like pliers." She went on to tell us about the most interesting discovery of all: these pseudowoodpeckers also behave as sapsuckers do on the mainland. A major part of Liba's field research was devoted to trying to understand why 'akis use "aki trees," or sap trees. Typically the birds forage on koa trees, but about 5 percent of the time they forage on 'ōhi'as, yet not for insects and grubs—they're after the sap, much like North American sapsuckers. "They drill into the surface of the bark and into the vessels in which the sap streams (the phloem). Then, while holding their slender upper mandible out of the way and using their lower mandible as a guide, they drink the sap with their tongue. You see them touch the tip of their bill or their tongue to the sap over and over."

Liba's eyes widened as she recounted her discovery of the secret life of her study bird: "If you stick an 'aki tree with a nail, as I did many times, the sap comes flowing out, and it is quite sweet." But the amazing thing was that she found the opposite to be true for the 'ōhi'a trees she tested that showed no evidence of previous hole drilling. When she poked those with a nail, no sap emerged.

"Somehow the birds know which trees are productive. Or perhaps they make them productive by changing the flow and pressure of the sap," Liba commented. Many of the 'aki trees that produce sap are covered in holes from roots to canopy. By following the

*'Akiapōlā'au (*Hemignathus munroi*) looking for grubs*

birds she had located, she found that each pair had as many as seven to ten sap trees in their particular home range. It turns out that only about one in a thousand 'ōhi'a trees is a sap-giving tree.

~

When it comes to rarities and even common species whose numbers are falling, the problem we face now is how fast the current activities of *Homo sapiens* are speeding up extinctions of everything Hawaiian, from land birds to snails to plants. The loss of birds, though, is the most obvious. "What really surprised me about the Hawaiian birds," Liba said, "was their susceptibility to the devastating impacts of invasive species by predation, competition, and disease." She sounded as if she were preparing an epitaph. "Few species have survived the plague that is avian malaria, the loss of key host plants such as lobeliads to pigs and other ungulates, and nest predation by rats and cats. I'm not sure there is anywhere else in the world where invasive species have had such a dramatic effect on a native avifauna."

Yet surprising data from another part of her 'aki study offered a glimmer of hope: "Remember the beautiful old-growth koa forest we walked through in Hakalau Forest? The one you thought looked like Middle Earth from *The Lord of the Rings*? When I started my 'aki project, most folks assumed 'akis were rare in part because they depended on big, old koas in places like Hakalau Forest." Most biologists were simply buying into the conventional wisdom of rarity theory: rare habitats should be where rarities are concentrated, and because there was little old growth left, the 'akis that depended on them were on their way out. Only it turned out that the 'akis didn't depend on the koa old growth exclusively.

Liba discovered that 'akis actually lived in higher densities in ten-year-old koa stands than they did in the old-growth koa forest of Hakalau Forest National Wildlife Refuge. She mapped home ranges at Hakalau Forest and on her study site covered by young koa trees at Keauhou Ranch, on the western side of the Big Island. "In the Hakalau refuge, 'akis have home ranges that roughly fit together like pieces in a jigsaw puzzle," she said, referring to the territories 'aki pairs had carved out for themselves. But she found the opposite to be true in the young stands at Keauhou: there, 'akis seemed not to be as territorial. Their home ranges overlapped almost entirely, with males defending females rather than chunks of forest. There are two parts to the explanation for this: first, she documented that 'akis spend about 90 percent of their time foraging on koa; second, she suggested that young, thick stands of koa offer a much higher density of food than do scattered giant koas in mature forests dominated by 'ōhi'a trees, such as Hakalau Forest's old growth. "In short," she concluded, "more food, more 'akis." Biologists working on other species, such as carnivores, show that territoriality becomes relaxed when the food supply is so abundant there is no need to defend parcels of land as hunting grounds. Such behavior would represent wasted energy, and there would be selection against it.

'Akis were fortunate that a real woodpecker or sapsucker never landed on the Big Island or was introduced by some misguided bird lover. Without a sapsucker around to compete with, the 'aki literally tapped into the most common tree on the island. Liba's findings also provided an important insight into how to conserve other rare species in a similar predicament. That 'akis seemed actually to be nesting and raising young in the even-aged koa forests with no difference in reproductive success was particularly exciting to her. "This means that perhaps we could plant koas elsewhere on the Big Island and both provide habitat for an endangered species and provide landowners with a tree 'trust fund' for their grandchildren." My puzzled expression indicated to her that I didn't immediately see the connections. I understood the first part: if 'akis and perhaps many other presumed old-growth-dependent bird species can reproduce well enough in immature forests, this is a measure of their resistance and resilience to extinction, and carefully designed forest regeneration programs could be truly beneficial to them.

Liba grew more animated. "Did you know that a koa dining room set goes for $12,000? And that individual koa-wood bowls turned on a lathe go for between $500 and $1,000?" Koa, Hawaii's premier hardwood, is referred to in some circles as "Hawaiian mahogany." The tree is also important culturally, and some people were now planting koas to grow canoe logs for traditional purposes. "Koa stands," she observed, "could provide a 'win-win-win' for saving the 'aki, bringing an economic return, and honoring Hawaii's traditional heritage." Provided, of course, that the koa seedlings are planted above the mosquito zone.

Liba's hopeful story contrasts with David Steadman's ecological history, which tells us that the prospects for many island bird species are grim. Steadman's long view concluded that more than 50 percent of Hawaii's endemic birds went extinct with the arrival of the Polynesians, another 20 percent had died off since 1825, and about 70 percent of the remaining birds were listed as endangered or threatened. Included on this list were eleven species determined to be basically unrecoverable or functionally extinct. Steadman summed it up to me this way: "The past few hundred years have only made the situation worse. While the surviving species of birds deserve our best efforts to save them from extinction, the sad truth is that there's not that much left to save."

Steadman is right in one sense: It is too late to safeguard most of these avian rarities because many of them are already extinct. The question that remains is how fruitful would be any effort put into saving the remaining species. And if we were to make a concerted try, what would be the best method? Recovery efforts now under way involve both species- and habitat-oriented approaches to conservation. Although there are many rare species worthy of recovery efforts, the Hawaiian bird species receiving the most attention so far, in addition to the 'aki, are the Hawaiian crow, the Hawaiian goose, and the palila.

The endemic Hawaiian crow, the 'alalā, was one of the rarest birds in the world when species recovery work began several decades ago. Biologists often joke that long after humans disappear and all the rare species before them, Earth will still be teeming with cockroaches and grasshoppers, with crows to gobble them up. Like other species of crows on Pacific islands, however, the Hawaiian crow is ecologically very different from our plentiful mainland crows. It is a forest bird, highly frugivorous, and it shuns agricultural and urban landscapes. "The 'alalā has declined for many of the same reasons as have other native Hawaiian birds—loss of habitat, hunting, disease, and introduced predators," Liba said. Before they can learn to fly, fledgling 'alalās are caught on the ground by cats and mongooses. Avian pox and malaria have also taken their toll.

By 2004, the last wild 'alalās were declared gone, so it is now up to the captive population to be the source of the recovery effort. The current 'alalā captive breeding program on the Big Island contains more than seventy-five individuals, but releases into the wild of captive-born young have been disappointing. Captive birds are treated for pox lesions before being released and their release area is beyond the malaria danger zone, but released birds are often killed by 'ios, native Hawaiian hawks. This circumstance is surprising because crows routinely mob and drive off hawks and eagles in other parts of the world. In Hawaii, even though the released birds seem to recognize hawks, they don't react. Experts have recommended removing Hawaiian hawks from the release sites by whatever means necessary, an ironic twist because the 'io itself is an endangered species (although currently being considered for delisting). Better proposals are to liberate the chicks on another island where the hawks are absent or to free them in dense forests where 'ios would be less common.

The prospects for the Hawaiian goose, the *nēnē*, is more optimistic. The nēnē looks far too tasty to have survived the Polynesian luau. A common phenomenon in native lands is the absence of rare vertebrates over the landscape—that is, until you hit pockets that turn out to be sacred areas where deities reside and humans dare not tread. The only reason the nēnē is still on the Big Island is that a remnant population of about thirty to fifty lived in the saddle between Mauna Kea and Mauna Loa, an area deemed kapu (taboo) and avoided by superstitious Hawaiians. Nevertheless, a concerted effort was required to coax the nēnē away from the gaping crater of extinction.

Today's stable population of almost 2,000 was achieved through captive breeding and release, the success of which depends on two factors: reintroduction into appropriate habitat and protection from predators. Adult nēnēs can fly, but they nest on the ground and the goslings are vulnerable for a long time. Soon after the recovery program began, it became clear that many birds released into marginal highland habitat that lacked abundant food plants subsequently perished. The nēnē's odds of survival in the wild picked up when mongooses, feral cats, and feral dogs were removed from the area. Now, with a better understanding of the habitat and protection needs of this species, releases are targeted more effectively. On Kauai, where mongooses never entered, predation is less severe and the nēnēs are finally doing well. Ironically, like Canada geese on the mainland, the nēnē is fond of golf courses, demonstrating an ecological flexibility to exploit and eventually rebound in human-altered habitats.

All approaches to recovering rare avian species and rare plants and insects on islands have similar requirements: remove nonnative plants and exotic mammalian herbivores and predators; fence off, if necessary, the critical habitat that can support the rare species; and let nature take its course. The idea is that rare native birds will respond to the positive changes under these management actions. In fact, Jack Jeffrey and his colleagues have shown that native bird numbers have increased at Hakalau Forest National Wildlife Refuge since forest restoration began twenty years ago. For example, 'i'iwis, 'apapanes, and Hawaii 'amakihis have more than doubled in density. The number of breeding birds in Hakalau's open forests has increased, as shown by the Hawaii 'elepaio (31 percent), 'aki (125 percent), and Hawaii creeper (39 percent).

In certain cases in which a species' population has fallen below one hundred individuals, captive breeding, reintroduction, and translocation have become important tools. But all of these interventions are very expensive, and conservation in Hawaii is underfunded. The Hawaiian Islands contain about 45 percent of the endangered species in the United States but receive less than 5 percent of federal funding.

Climate change has emerged as a long-term threat to the persistence of rarities in many places we have visited, and Hawaii is no different. Besides the issue of birds moving upslope to find the microclimates they prefer, their parasites might tag along with them in a warming climate. In the 1990s, the fear was the introduction and establishment of a species of *Culex* mosquito that could survive the cool temperatures of the montane forests and spread deadly disease in the presently malaria-free zone. However, Peter Vitousek informed me, "it's too cold for malaria to complete its life cycle in time, even though a competent vector—this new species of mosquito—is there." But if temperatures warm up the mountainsides, the malarial parasites could survive and infect the birds that used to live above the zone.

New studies show that at least one of the native forest birds is holding its own with the local strain of avian malaria. A 2007 paper by Bethany Woodworth and colleagues documented, for the first time, that the 'amakihi seems to be developing some resistance to malaria. Another, yet untested hypothesis is that the disease is becoming less virulent, a phenomenon of natural selection that leads some variants of the pathogen to persist and continue to infect the host. This finding was the result of a large project about ten years ago that found, quite unexpectedly, that 'amakihis were abundant at some low-elevation sites. Before this study, the conventional wisdom was that most native birds could not really persist below the "disease line" due to avian malaria. 'Apapanes also seem to be increasing in abundance at low elevations. The fecund 'apapane, which can have multiple broods of two or three chicks in a season—as compared with the 'aki, which typically has one at best—may be evolving faster to adapt to surviving the disease. Extinction may lurk in the shadow of rarity, but natural resistance to avian malaria, as is being seen in some species, gives more hope for growing resilience in others, if it is selected for soon enough. That is a lot to hope for.

On one of our last nights in Hawaii in 2001, David Wilcove and I approached the active flows of Kilauea to watch the stream of lava rolling into the sea. Under a full moon, we witnessed Kilauea in its fiery adolescence. We hopped over the black, twisted ropes of long-cooled *pāhoehoe* lava. There was nothing else—everything had been incinerated or buried in its wake. Many of the tourists scrambling about the lava field, headed for the same destination, were here for the spectacle—the sight of pieces of lava like fiery ingots dropping into a cooling sea. A few biologists were also about, snapping photographs, probably for future lectures on plant succession. In a crack in the lava bed, the presumed lecturers gathered around a pioneer—an *ʻōhelo*, a kind of huckleberry—and nearby was a *kūkaenēnē*, whose fruit is a mainstay of the Hawaiian goose. These plants were establishing themselves in a barren landscape, giving organic life a toehold for expansion.

On volcanic islands in the species-rich tropics, a single eruption can wipe out millions of years of evolution in one moment of frenzy. But without lava flows, there would be no Hawaiian Islands. The hillside above us bore the scars of old lava streams, some ancient and long since revegetated, while others were still fresh and simmering. We have no control over the timetable for the next eruption of Kilauea, or Mauna Loa, or Mauna Kea. But we have considerable leverage over the timetable for the fate of the remaining honeycreepers, land snails, and lobelias and for the rest of wild Hawaii. It is in our hands, not just at the mercy of the gods of the mountains.

Chapter 8

Ghosts of Indochina

*M*ANY OF THE WORLD'S most popular wildlife view-
ing spots feature tame, but still rare, species that tolerate human
proximity. The gray whales in Mexican lagoons, inquisitive giant
tortoises on the Galápagos Islands, and habituated mountain goril-
las in Rwanda are all comfortable in our presence, failing to link
humans with our penchant for violence. Such innocence vanishes
in the wake of war. On former human battlegrounds, wild animals
flee at the snap of a twig or the first scent of any human intruder.
They sink into the foliage, becoming ghosts to those who search
for them. Once they have experienced the chaotic fear of war, most
species retain their dread of humans even long after hostilities have
ended.

Nepal has suffered from a Maoist insurgency and the battles
against it, and Peru struggled through the Shining Path era. Neither, however, faced the seemingly endless conflict waged in two

other hotbeds of rarity, Vietnam and Cambodia. For a good part of the last half of the twentieth century, the forests and wildlife of Indochina endured invasion, carpet bombing, napalm, unexploded ordnance, land mines, defoliation by Agent Orange, and a heavily armed, protein-starved populace. For fifty years, naturally rare species clung where they could to survival in the midst of mayhem.

The United States' war in Vietnam ended in 1975, but the Indochina region remained closed to Western biologists for at least ten years longer. By then, the economies of Vietnam and Cambodia had begun to recover from decades of war and deprivation. But many biologists had spent the decade wondering about the fate of the irreplaceable wildlife that once flourished in the rich mountain forests, river deltas, mangroves, and coral reefs of Indochina. After the fall of the Khmer Rouge in Cambodia in 1979, was the unusual woodland wilderness east of Angkor Wat in recovery, too?

The impact of war on a fragile peninsula such as Indochina is by no means unique. Many other biologically and rarity-rich regions have suffered a similar fate. A short list includes Angola, Mozambique, South Sudan, Rwanda, the Democratic Republic of the Congo, Afghanistan, and Korea, but one could add at least twenty other examples.

This sad but all-too-common phenomenon of human conflict is the last major human-induced cause of rarity for us to examine up close. An important question to answer underpins what at times resembles a search-and-rescue mission for rarities: Do the outbreak and aftermath of war inexorably drive both rarities and some former commoners to extinction? Our first stop is an exemplary destination, the Annamite Mountains, known in Vietnamese as the Truong Son. In any language, this is one of the most biologically unexplored cordilleras on Earth, a necklace of remote ridges along the western border of Vietnam that are studded with the highest concentrations of rarities in mainland Asia. Neighbor to the Annamites are the Eastern Plains of Cambodia, a vast region bordering Laos and Vietnam, formerly home to elephants, primates, tigers,

Map of Vietnam, Cambodia, and the Annamite (Truong Son) range

rare giant cattle, and other large ungulates. This is a story of the rehabilitation efforts, from the early stages beginning in 1985 up through July 2012, to bring rarities back from a war-ravaged land, once peace had returned.

∼

The first Western biologist to reach postwar Vietnam in the mid-1980s was John MacKinnon, Southeast Asia's most acclaimed field man. The Oxford-educated MacKinnon was England's answer to Indiana Jones, a superb biologist but at heart a devoted rarity hunter. His impressive body of work includes a guide to the birds of China; books on the birds of Sumatra, Borneo, Java, and Bali; and his most comprehensive opus, a review of protected areas in the Indo-Malayan faunal realm, a vast region stretching from Pakistan to the eastern edge of Indonesia. Many of the park systems of Southeast Asia have John's fingerprints on their design. His guiding principle: it is crucial to protect examples of all the habitat types in a country so as to conserve the widest array of species. To help develop Vietnam's new protected-area plan, John had received permission to search every habitat of this long, narrow nation lying between the Annamites and the South China Sea. At a time when most Westerners were prohibited from exploring zones most directly affected by the fighting, MacKinnon's freedom was envied by the global community of biologists.

The next arrival, in May 1990, was a younger conservator of rarities. Twenty-six-year-old David Hulse was responsible for fashioning the training program for future staff of the new park system being designed by John and his Vietnamese counterparts. David's party included Bernard Masters, an Australian education specialist with the World Bank, and me. I was new to fieldwork in Indochina, as was just about everyone but MacKinnon, but I had spent a few years in the region as the World Wildlife Fund's Asian conservation scientist. Our field visits would be limited to reserves near Hanoi, far from the places bio-explorers such as MacKinnon planned to go. Still, on this trip we would at least be able to interview Vietnamese biologists searching for the wild orphans of war.

During our flight, David had been going through an English-language version of a Vietnamese newspaper he had found at the Bangkok airport. As our plane approached Hanoi, an unusual story caught his eye. He examined the grainy photograph of Asian elephants immersed in a jungle wallow and handed it to me. It was striking not because of the behavior captured in the image—large tropical mammals frequently slip into wallows to prevent heat stress. Rather, it was the symbolic nature of the makeshift wading pool: it was a crater left from a B-52 bombing raid thirty-five years earlier.

Before the trip, I had visited the Smithsonian Libraries to gather what information I could about rare Vietnamese species. The darkened stacks held one obscure book, *Three Kingdoms of Indo-China*. Published in 1933, it described a natural history expedition led in 1927 by the late Harold J. Coolidge Jr., a Harvard mammalogist, and Theodore Roosevelt. Their journey along the Mekong River and its tributaries cataloged the biology of a region covered in a brocade of tropical forests, remote mountains, and secluded valleys that supported unusual animals found nowhere else in the world. The expedition collected many small vertebrates, but the biologists were most interested in the four large mammal species—the elephant, a wild water buffalo, a wild ox called the banteng, and another called the gaur, also known as the seladang or Indian bison.

Other nuggets on Vietnamese rarities emerged with a bit more detective work. I read, for example, of a wild animal shipped in 1937 from Saigon to the Vincennes Zoo in Paris that turned out to be not the calf of a gaur, as assumed, but something entirely new to Western science. As a result, the regal-looking kouprey, a gray ox that roamed in small herds across the savannas and woodlands of Indochina, was first described. Larger than Brahma bulls, at 900 kilograms the males carried such an impressive rack that they could be called the Cambodian longhorn. Their magnificent horns often sported a ring of abraded splinters known as a "crown of thorns," the result of the bulls' frequent digging into the earth with their pointed headgear. The reason for this behavior is unclear, but it might have been a form of dominance display. The captive kouprey in Vincennes never had the chance to show off its imposing profile, though. It lived for only three years before starving to death during the German occupation of France in World War II.

Another discovery was a monograph published in 1957 by biologist Charles H. Wharton, who led an expedition to Cambodia and became the first to film wild koupreys. His planned decade-long field study ended prematurely when the war of independence against the French broke out in the early 1950s. What we know about this creature in the wild comes largely from Wharton. For decades, conservationists ranked the kouprey among the thirty rarest mammals in the world. Although few people read Wharton's monograph, many saw his remarkable 1957 film about his Cambodian journeys. A wildlife classic, the prosaically titled *Forest Cattle Survey Expedition to Southeast Asia* is a tour de force among nature documentaries. The film describes the eastern plains of Cambodia as Asia's Serengeti, harboring high densities of large mammals. Some of the scenes are priceless: Wharton, bare-chested at the wheel of a Willys Jeep bouncing through malarial jungles; sharp-featured guides using cigarette smoke to test the wind while tracking big game; and the images of large mammals Wharton captured for posterity on film—gaurs, bantengs, wild water buffalo, Eld's deer, and koupreys.

All of these pieces of biological history replayed in my mind during the two-hour flight from Bangkok to Hanoi. I returned to reality when, with a sharp bounce, our plane hit the tarmac. Minutes later, we were inside the terminal and soon we were through customs. A young man sporting a Beatles-like mop of hair held up a sign with our names on it. Although his attire suggested Asian disco—a flashy safari suit and shiny black shoes with platform heels—he was our chaperone and translator, Mr. Tam, now smiling broadly and waving. "Welcome to Vietnam," he offered in excellent English.

*The elusive, probably extinct, kouprey (*Bos sauveli*)*

After exchanging pleasantries, we piled into a waiting van, which turned out to be one of the few cars in the city at the time. On the crowded thoroughfare to town, hordes of bicyclists swerved around us at intersections. We turned off the main road and the scene instantly changed. French colonial villas bordered wide, quiet boulevards shaded by overhanging trees. This charming historic district had somehow survived bombardment. I briefly entertained the image of relaxing on an elegant balcony draped with showy blue-flowered *Thunbergia* vines, jasmine, and multicolored sprays of bougainvillea. Instead, we arrived at the Ministry of Defense Guest House. Our meager accommodations would never grace *Fodor's Vietnam*, but they were close to the Ministry of Forestry, where we would be working. The guesthouse cafeteria provided our daily rations: fried spring rolls and *pho*, the famed noodle soup of Vietnam. Female soldiers served us at the guesthouse. They smiled but kept a watchful eye on perhaps the first Americans they had seen since the war ended.

The next morning we dressed in coat and tie and set off to meet ministry officials to obtain approval for our proposed conservation program and the necessary permit to travel outside of Hanoi. We entered a large room crowned by a reluctant ceiling fan that, even at full throttle, was no match for the humidity of early May. Bernard, David, Mr. Tam, another interpreter, and I sat on one side of the long rectangular table, and the department heads sat across from us. After three days of meetings, we finally won permission to start our work, and we were itching to talk to local scientists and conservationists.

Our first appointment was with Professor Pham Mong Giao, who had led expeditions to areas that were off-limits to us, where it might be possible to see the Javan rhino. A remnant population of fewer than a dozen individuals had somehow survived the hostilities by hiding in dense forests defended by the thick canebrakes of Cat Loc, close to Ho Chi Minh City, the former Saigon. "I have found tracks of the Javan rhino," he told us excitedly, "and dung piles, but unfortunately, no sightings." Weighing about 1,000 kilograms, Javan rhinos are perhaps the sixth-largest living land mammal, but spotting one, like spotting a kouprey, seemed improbable.

How circumstances had changed for this now rare megaherbivore. Until several hundred years ago, the Javan rhino ranged far from its namesake island all the way to northeastern India. Biologist George Schaller once told me that, surprisingly, before the French Indochina War in the 1950s more Javan rhinos may have roamed in Indochina than on Java.

As Professor Giao explained more about his field studies, it occurred to me that Javan rhinos possessed a trait that had served the species well during wartime. Their ability to adapt their behavior from a diurnal feeding cycle to one that was nocturnal and silent equipped them to take shelter in the daytime. They survived on the browse available while hiding in an impenetrable thicket of spiny rattan that even guerillas shunned. As long as the Javan rhino's energetic requirements could be met by eating only at night, it could avoid being seen.

We made plans to meet John MacKinnon, who was also in town, at a cafeteria favored by university students. It wasn't hard to spot our man—a tall, pale Westerner with gray hair, the only person there over the age of fifty. He would be headed next, he said, to a mountainous section of the Vu Quang area northwest of Hue. "Satellite photos show this to be the largest block of intact forest in the country," he noted. "It's very wet, it's a highly restricted area, and no Westerner has been in there." Biologists have long suspected that the wettest tropical forests hold the greatest hotbeds of rarities. The scientific rationale for the rarity distribution: superwet forests probably served as refuges for numerous species during dry epochs, and when the wet periods returned, many stayed put.

John's brilliance far overshadowed his reputation for being difficult. In addition to helping to set up Vietnam's new park system, he was also here to find koupreys and other rare large mammals rumored to be lurking about. Drawn into a tense competition with several Western biologists to find these missing species, he planned to reach them first.

After meeting with John, we headed to Cuc Phuong National Park, Vietnam's answer to Yellowstone. Reaching the park would require driving for five hours on a rough track and ferrying across several rivers. During our first hour or so out of Hanoi, the landscape revealed little of interest. As we drove on, the eroded limestone cliffs we saw rising from the plains resembled the foreground of an ancient Asian landscape painting. A copse of lowland forest was hard to find in the vast panorama of paddy fields and limestone pinnacles between Hanoi and Cuc Phuong. Vietnam was well on its way to challenging Thailand as the world's number one exporter of rice.

Sadly, our tight schedule and government chaperone gave us no time to stop and explore one of the world's greatest yet least known reservoirs of rarities. Had we ascended the limestone cliffs, we would have found enough specimens on their rich flanks to spend years cataloging an impressive new collection. When Tony Whitten, an expert on animals in limestone habitats, surveyed land snails along our route and in northern Vietnam, he discovered that almost one-third of the 270 species he encountered were new to science. Besides land snails, many species of plants and insects are also limestone specialists; searching a previously unexplored outcrop can yield a haul of invertebrates of which 90 percent will be new to science. Ha Long Bay, made famous in the film *Indochine*, is studded with more than 2,000 breathtakingly sculpted limestone islands, many holding rare orchids, balsams, and begonias found nowhere else. The limestone bluffs in Vietnam offered in miniature a pattern that has emerged across many limestone-rich areas in the tropics: hotbeds of species with extremely small ranges on unusual soil types that have been reproductively isolated for long enough periods to allow for speciation. Tragically, we observed dump trucks backing up to the base of a limestone tower. The rarity-rich mountain of accessible limestone was being mined for building materials. It was the first visual sign of our journey that nature conservation would take a backseat to building the new Vietnam.

It wasn't until we entered the buffer zone of Cuc Phuong that the rich Vietnamese forest appeared. At the headquarters we met the park director, Nguyen Ba Thu, a gentle man in his late forties. He gave us a brief introduction to the park, and we then spent the afternoon poking around the forest looking for some of Cuc Phuong's rare primates, among them the Delacour's langur, a highly threatened species native to the park. Of the twenty-one species of primates found in Vietnam, fourteen are endangered. Many have startlingly human faces, especially the bewhiskered red-shanked douc, whose eyes are shadowed in powder blue.

The next day, Mr. Thu took us on a walking tour of the park. Besides its langurs, Cuc Phuong is famous for a mammal endemic to Indochina, the Owston's palm civet. Civets resemble long-snouted weasels with bold stripes. This particular civet is restricted to Laos, northern and central Vietnam, and nearby China. Cuc Phuong is one of the few parks that offer this civet protection so it can pursue its favorite activity—hunting night crawlers—without risking capture and ending up on a dinner plate. Palm civets are also expert coffee pickers. Civets roam freely about the plantations and harvest only the ripest berries. Their droppings contain select beans that, when picked out, washed, dried, and roasted, sell for a fortune. In Vietnam these beans are marketed as "weasel coffee," in Sumatra as *kopi luak* or *kopi musang*, "civet coffee." Either way, some insist civet droppings yield the world's finest java.

But after an entire day of exploration, we had seen neither civet nor langur. Our mammal list had not expanded beyond a few squirrels. We had encountered leeches in great abundance, however. Terrestrial leeches are rare in the New World (although there is a rich diversity of aquatic leeches) but ubiquitous in Asia above 2,000 meters and in moist lowland habitats. One explanation for the dearth of leeches in the New World is that vast blood reserves disappeared when the region's large mammals went extinct in the Pleistocene epoch. In contrast, Asia is still a leech outpost, with plenty of large mammals on the menu, including humans.

Judging by our experience, however, leeches in Cuc Phuong must have had a hard time finding those mammals. Excessive hunting had turned this area into a virtual "empty forest"—a forest without its vertebrates—an increasingly common sight in the tropics by the 1990s. One cause was certainly the pressures exerted by specialized poachers on the persistence of rarities—from jaguar trappers to rhino horn thieves to smugglers of tiger bone and elephant ivory—and feather hunters, as we have seen in New Guinea, Peru, Nepal, Brazil, and Hawaii.

But another group is an even more disastrous contributor to the empty forest syndrome: bushmeat hunters. The term "bushmeat" refers to any wild game caught by locals. Sometimes the meat is for the family pot, but more often and more damaging, commercial hunters capture or kill wild game for restaurants or wealthy urban customers. Commercial bushmeat hunting is particularly pervasive in the tropics. There is practically a subdiscipline in the field of conservation biology devoted to describing its staggering footprint and how to contain it. We began to suspect that the communes around Cuc Phuong were home to among the most effective bush-meat hunters anywhere, as the forest surrounding us remained ee-rily silent. What we were witnessing was in microcosm a major challenge facing many former war zones: the soldiers return home with their weapons and the skills to use them.

Two days later, we were back in Hanoi with John MacKinnon to attend an all-day symposium on forests and conservation. We were the only Westerners present. At lunch, we were ushered to a seat next to a distinguished-looking elderly man in uniform, who turned out to be General Vo Nguyen Giap, a famous Vietnamese soldier-turned-conservationist. I knew of his reputation as Viet-nam's Eisenhower and as a superb tactician, but I wondered what made the general a conservationist. In his fluent French, he ex-plained, "The forest is our friend. It hid us and provided shelter, food, and water. Now it is our turn to save it." The forested areas at the borders with Laos and Cambodia had allowed North Viet-namese soldiers to infiltrate the south and conceal their divisions along the Ho Chi Minh Trail. The dense Vietnamese forest hid more than soldiers and material, though. It had also become the last sanctuary for the region's forest-dwelling rarities.

∿

In 1992, two years after that initial trip, David Hulse was among the first conservationists to return under license from a US govern-ment program seeking to build relationships with Vietnam. Along with him, a small army of Western naturalists were eager to explore the Annamites and beyond. MacKinnon would have to hurry if he wanted to beat the pack in sighting the kouprey and other lost species.

There was a silver lining to such earlier unsuccessful kouprey surveys and related searches in the secluded Annamite range. Field scientists had uncovered a seam of rare species that startled the world, reporting new finds across a wide range of taxa: a bovid, a rabbit, two deer, several birds, and hundreds of fish. It was as if, eons ago, the Asian plate had tilted east, and all the rarities had rolled toward the South China Sea before dropping into hidden valleys of the Annamites.

Meanwhile, MacKinnon continued his quest for the kouprey. "It's a bit like looking for the Yeti or Bigfoot, this animal," he told a reporter who had come to cover his story. "First, it was just extremely rare and then it was shrouded in mystery through 30 years of warfare. It's become sort of a symbol of conservation in Indochina." After extensive surveys in the company of Professor Giao and other Vietnamese field men, MacKinnon's team struck gold that year, though not in the form of his original quest species. During an expedition to the Vu Quang Nature Reserve in Ha Tinh Province, they were invited into some hunters' houses. The hunters placed three skulls on the table of something entirely new to the Westerners. The sharp, saber-like horns resembled the weapons of Africa's scimitar-horned oryx. And so *Pseudoryx*, or, as it became known, the saola, entered the annals of the mammal kingdom.

News of the saola spread fast. Further surveys of the Vu Quang region turned up more than twenty specimens, and at last a wild saola was caught in 1994. An account in the journal *Nature* raised its profile. David Hulse made his own trip to the area, and his stunning photographs of this peculiar ungulate gave a face to the name. All of this attention raised an obvious question. How could a large herbivore weighing more than 90 kilograms and with formidable horns stay hidden from scientists, Vietnamese or Western, for so long? The answer was that the saola, as the Javan rhino once did in its own range, lived deep in the forests of Vietnam and Laos along the Annamite chain. The area was wet and malarial, had poor soils, and had a very low human population in an otherwise densely populated country. And saolas seemed to survive best in the higher steep valleys, where few locals lived or hunted. By heading straight for this dense, wet region of forest that Coolidge and Roosevelt had circumvented in their 1927 expedition, MacKinnon had seen his

*Saola (*Pseudoryx nghetinhensis*), the most recently discovered large mammal*

hunch pay off. The best place to find rare species during postwar recovery was in the biggest, wettest block of unsurveyed intact forest. The saola's success at adapting shows another coping mechanism. Selection for being a habitat generalist when a species first evolved, long before the human era, allows it to persist in remote habitats on poor soils, on ground too nutrient poor and too steep to grow crops. In Vietnam and elsewhere, this becomes the definition of a modern refuge.

MacKinnon never found a kouprey in the wild, nor has anyone since. Soon after, though, members of his intrepid field team announced the discovery of another new species, the large-antlered muntjac, a member of the ungulate group known as barking deer. Also known as the giant muntjac, the thirty- to fifty-kilogram reddish-brown deer is considered endangered and is known from only a few locales. The next few years of searching yielded another muntjac, this one named the Annamite muntjac. The taxonomy of muntjacs remains shaky, and further genetic analyses may reveal four or more species in the Annamites alone. These were the first new large mammals identified by Western scientists in Indochina since 1936, when the kouprey was first described. This rich vein of new species was hardly what the French naturalist Georges Cuvier had in mind when he proclaimed in 1812 that "there is little hope of discovering new species of quadrupeds."

In the past decade, Western biologists have come to recognize what great natural wealth is packed into the Annamite Mountains, the rest of Vietnam, and neighboring Laos, and they have been on a quest to explore every secluded valley and mountaintop in the region. The windfall of endemics and newly discovered species made some scientists wonder how so many early naturalists could have missed so many novelties. One explanation is that many of these habitats had been too dangerous for outsiders to visit. Local people repelled visitors, often violently. Deadly malaria in the remote and rainy Annamites was also a serious problem. Finally, the Annamites were so steep, and walking along their sharp limestone trails so challenging, that early naturalists were more attracted to the deltas and flat terrain full of birds and plants.

The cryptic nature of the saola and the muntjacs—and, by extension, of other small, territorial forest-dwelling Asian and African ungulates, such as duikers (tropical Africa), chevrotains (South and Southeast Asia), and musk deer (Himalayan regions)—may be the result of another strong selection pressure. The ability to avoid detection by prowling leopards would be a behavioral trait that natural selection would reward. So perhaps a feature that was selected for eons ago serves well today. A muntjac or saola skilled at evading detection by leopards must be hard to find for predatory humans, too.

As was true in New Guinea, understanding a bit of the area's geologic history can explain a lot about the current concentrations of rarities in Indochina. About 350 million years ago, the Annamites arose during collisions of Earth's plates. They are truly ancient mountains—much, much older, for example, than the Nepal Himalayas, at about 55 million years old. The Himalayas emerged when what is now India broke off from the ancient supercontinent known as Gondwanaland and rammed into Eurasia. Initially, the joining of the subcontinent of India with Eurasia opened a vast new landmass for the invasion of novel species. Then the uplift of the Himalayas and the Tibetan Plateau sealed off Southeast Asia from the northern part of the continent, preventing new invasions. This event isolated the Annamite fauna and flora.

Other mountain chains arose in the northwest of Vietnam, followed by the emergence of highlands in the northeast. Changes in river courses and, along the coasts, the rise and fall of sea levels successively isolated and rejoined populations of plants and animals. When sea levels were low, connection by land bridges to the south allowed an Indonesian influence to reach the flora and fauna of Indochina. In addition, cyclic weather patterns featuring winter monsoons and summer rains drenched some areas and passed over others.

In sum, we can see in Indochina all the ingredients for a rarity stew. Ancient mountain range developments and shifting patterns of water and land promoted conditions of physical isolation conducive to speciation, as did variegated patterns of warm temperature and precipitation. The narrow bands of wet and dry forest resulted in a rich mosaic, offering a diverse range of habitats where species could settle and might remain in isolation from one another.

~

The search for further evidence of the Javan rhino in Indochina picked up in earnest in 2003, when an energetic Dutch biologist, Gert Polet, began supervising fieldwork in Cat Tien National Park. His team looked for rhino spoor, any sign that this elusive creature persisted in the former battleground. Cat Tien, about 100 kilometers south of the terminus of the Ho Chi Minh Trail, had been an important staging area for soldiers going into the south. US and South Vietnamese regulars came here to stop them. There is no record of any encounter with Javan rhinos during the battles, and Polet's team found no evidence on the ground.

Polet's team's use of infrared camera traps had an important precedent in an attempt to catch the phantom mammal on film. One of these traps, set in 1998 by WWF scientist Mike Baltzer, captured on film a Javan rhino reaching up to nibble a tree branch. The image was seen around the world, but depressing news accompanied it: Gert and Mike's final surveys in 2006 put the total number of Javan rhinos at fewer than a dozen in Cat Loc. Interestingly, Cat Loc, a forested area with no formal protection, is very close to Cat Tien National Park. A dense, thorny thicket of spiny palm draped the hilly terrain of Cat Loc. Professor Giao's rhinos had become ghosts within it. Of concern was that no one had seen small footprints signifying a calf. Biologists sought explanations for the population's perilous drop from the dozen or more thought to still be resident in Cat Loc. One immediate speculation was that they had been poached for their valued horn. Another was collateral damage from the bombardment some years earlier. But in order to understand what happened to them and the other rare large mammals, we need to revisit the war years and make inferences based on a bit of human epidemiology.

Defoliant. The soft, gentle sound of this noun suggests a home beauty product rather than a deadly toxin that kills tree leaves. As General Giap mentioned over lunch in Hanoi during my first trip to Vietnam, he and his troops used the forest strategically and hid within it; defoliant was designed to expose them. The American strategy from the mid-1960s to 1971 included an unprecedented spraying campaign intended to halt the infiltration of soldiers from the north and shift the course of the war. By 1971, more than 75 million liters of a defoliant dubbed Agent Orange (from the identifying orange stripe on the steel storage containers) had descended on about 57,000 square kilometers of inland tropical forest, an area about twice the size of the state of Massachusetts. Almost 1,300 square kilometers of coastal mangroves were also affected.

Agent Orange did much more than strip leaves from trees. Dioxin, a chemical generated during the synthesis of the growth hormone forming the defoliant's major ingredient, turned out to be highly carcinogenic. Soldiers on both sides of the conflict absorbed Agent Orange and stored it in their tissues. Because dioxin remains stable for decades, it is a persistent public health risk. Worse, it is able to cross the placental wall into the embryo, so postwar offspring were often affected.

Exposure to dioxin is linked to at least twenty-eight serious diseases, including a fatal type of leukemia. Because symptoms can take years to develop, government officials at first dismissed the claims of veterans who reported severe health problems from exposure. An out-of-court settlement between the companies producing Agent Orange and the 2.4 million Vietnam veterans filing suit against them led to a $180 million fund to pay veterans' health claims. The fund was quickly exhausted, however. A 2003 US Supreme Court decision enabled many more veterans to receive treatment.

So little was known about the effects of Agent Orange on wildlife that a survey was called for, to the displeasure of the US government. Among the first to investigate the ecological effects of the war was one of the leading ecologists of our day, Gordon Orians. Traveling as a correspondent for *Science* magazine, Orians sought to uncover the lasting effects of defoliant on nonhumans and particularly on rare species already at risk of extinction. In 1970, Orians and his colleague E. W. Pfeiffer published a paper in *Science* titled "Ecological Effects of the War in Vietnam." The defoliated upland areas he saw had been converted not to grasslands but rather to "bamboo forests." Bamboo, a grass, is not killed by Agent Orange. (The US Army used a different defoliant to target grasses in its "Resource Denial Program," which focused on spraying rice fields to starve the enemy and which could have affected the bamboo groves as well.) This defoliation of forests must have displaced species that depended on forest cover. Quantitative data on forest wildlife during the war were absent; there was one grisly anecdote, though. Unlike other rare vertebrates that fled battlegrounds, wild tigers headed toward gunfire. They had learned that combat zones meant encounters with mortally wounded and dead soldiers.

For the Javan rhinos in Vietnam, the best short-term strategy for preserving the species after the war was to hold on to the rhinos that had survived; the best long-term strategy combined vigilance and patience. As a first step, David and his Vietnamese colleagues worked through the 1990s to build up Cat Tien National Park through Dutch funding and create a conservation program in Cat Loc, where the rhinos actually lived. The second step was to design a viable future for the species by promoting conservation-friendly activities in a 2,500-square-kilometer buffer zone between Cat Tien and Cat Loc and to lay the groundwork for turning the entire area into a biosphere reserve under the auspices of UNESCO's Man and the Biosphere Programme.

~

In 2005, fifteen years after our first visit, David Hulse returned to assess the progress of Vietnam's recovery of its rare species. David invited me to accompany him to see one of his favorite conservation projects in the southern part of the country. Nguyen Tran Vy, a former protégé and now one of the top young ornithologists in Asia, was our host, and he was eager to show us some of the extraordinary birds in the Tan Phu State Forest Reserve in southern Vietnam.

On a hot, muggy November morning, Vy motioned for us to take cover behind a copse of bamboo. He pressed the "play" button on his recorder and the rising whistle of an orange-necked partridge arced over the forest. From the shoulder of the hill came a return volley. This was the fifteenth bird to answer his playback, but none had dared to come into our view. David and I stared hopefully at a gap near the forest floor, waiting, listening, sweating, while Vy's tape let loose more whistles. We hid, drenched in perspiration, blind to the male partridge examining us from a termite mound right next to us. Only our young guide saw the bird before it vanished.

For the past decade, Vy had kept watch over this bobwhite-sized hermit, known only from a few hillsides northwest of Ho Chi Minh City. But even if we didn't glimpse a single partridge after four hours of searching, we had encountered a multitude of vigorous singers—babblers, barbets, drongos, tailorbirds, flycatchers, orioles, shamas, trogons, and even rare Siamese fireback pheasants—all survivors of the shock of war.

We had arrived in Vietnam this time on Ho Chi Minh's birthday, May 19. Billboards with Ho's image were posted around the city renamed in his honor, exhorting the masses to make new sacrifices for the good of the motherland. The Communist-style message boards stood in stark contrast to the sights and sounds of roaring capitalism. Alongside the road into town stood a life-size statue of Colonel Sanders, taller than most of the Ho Chi Minh statues I noticed, welcoming patrons into a packed KFC franchise.

The trajectory of the rebounding economy was quite different from that of wildlife recovery. Reports to David from Cat Tien suggested that, by the new millennium, the number of Javan rhinos there was still dangerously low. According to local rumor, only four adults remained—all females. We would get a firsthand report from Cat Tien, which was our next destination after this interlude in Tan Phu.

To crouch motionless in the moist heat and mud seemed more like jungle warfare training than birding. Ironically, this former combat zone had recently become an important conservation area. Vy's surveys had found the highest recorded density anywhere of orange-necked partridges here in the forest reserve, and, thanks to his efforts, logging in the area had ceased. Beyond Vietnam, this shy bird had made international headlines by joining the list of the world's rarest vertebrates as an Alliance for Zero Extinction species (chapter 1). And this site was listed as the only place where it was found.

Between the pesky mosquitoes, the dense heat and humidity, the mud, and the leeches, that day in Tan Phu was about the most uncomfortable I had ever experienced in the field. Yet the more difficult the terrain became, and as the trail seemed to disappear into the sharp-stemmed bamboo, the happier Vy seemed. He noticed the mosquitoes preferentially attacking David and me despite our superstrength US-purchased repellent. "Try this one," he said, handing us a Vietnamese knockoff of Avon Skin So Soft that proved amazingly effective. I didn't want to ask what was in it but was thankful it worked. This part of Vietnam still suffers from a mosquito-borne cerebral malaria.

By afternoon, after many near observations of the reluctant partridge, our stomachs were growling for our adopted staple of pho, so we headed back into town. As we walked to the open-air restaurant, locals stopped to greet Vy. We were traveling with a celebrity, it seemed. Vy had his own television show for children to nurture their interest in Vietnamese nature; he knew as well as anyone that the next generation would need passion and dedication to carry on the work of saving the unique species of the country.

At the restaurant, we chose a table directly across from a countertop laden with large jars of seahorses, insects, lizards, and flowers suspended in rice-based alcohol. These were specialty cocktails thought to be infused by the essences and powers of the creatures within. One vat filled with baby king cobras sported a name on the label: "One Night, Five Times." Perhaps here was an insight, in the bottle, that locally held perceptions, however misguided, trumped scientific rationales. Vietnam was rapidly becoming the end point for Asia's rarities, which ended up in a bottle or as a cream or powder. If the mind-set here is to change, it will be people such as Vy, dedicated local scientists who can connect with a larger audience in their native language, who will lead the way.

The main purpose of our visit was to check on the rhinos, so the next day we headed with Vy to Cat Tien National Park. In the village of Cat Tien, we drove to the edge of the Dong Nai River and waited for a barge to ferry us across. On the way there, we had traversed part of what was seen as the "buffer zone" between Cat Tien and Cat Loc. It offered little cover for a passing rhino; that much was clear. In fact, in this section the buffer zone was a string of villages with hardly a tree in sight. Progress was being made on paper in establishing a legal buffer zone, but the effects on wildlife restoration seemed invisible.

Once across the river and inside the park, however, we found ourselves in a green wall of thick forest exploding with the sounds of birds. Several species of kingfishers wailed from the riverbanks; parrots and parakeets flew overhead; hornbills announced their arrival in the fig trees. By the end of our morning's walk, we had recorded nearly seventy-five bird species, thanks largely to Vy, who instantly recognized the songs of whatever was hiding in the tangles of vines ahead of us. He pointed out a scaly-breasted partridge walking along the forest floor. Other avian highlights in this former battle zone were an Asian paradise flycatcher and a host of black-and-red broadbills. That the words "paradise" and "former battle zone" could be used in the same sentence is a testament to the powers of nature to recover. Paradise flycatchers are exquisite birds, with males bearing long white tails that flutter like ribbons when they fly through the trees. Common across Asia, they sit on tiny cup nests that seem much too small to support the female and her eggs. The black-and-red broadbill, which boasts a pattern of bright crimson and a broad bill the aqua-blue color of a robin's egg, made me think of a small crow that had agreed to a total makeover.

At lunch, we resumed a discussion we'd been having of the Javan rhino. The Cat Loc population was already what biologists term "ecologically extinct" because they no longer played their natural role as landscape engineers. They were also, it seemed, about to pass from ecological extinction to true oblivion. A last-ditch proposal to keep that from happening involved an influx of new blood from Indonesia. If it turned out that the remaining individuals in Cat Loc were all female, it might be possible for the Vietnamese to negotiate with the Indonesian government for stud service or even permanent residence of several males and females that could be shifted from Ujung Kulon National Park in western Java to Cat Tien. However, the chances of a successful agreement were slim. For the past twenty years, conservationists had been unable to persuade the Indonesian government to move rhinos from western Java to two other parks just across the narrow strait that separated Ujung Kulon from southern Sumatra. Moreover, opponents of the proposed transfer from Java to Vietnam had argued that the Javan population is a different subspecies and should remain separate from the Indochinese variety.

Nevertheless, moving some animals from Ujung Kulon to Vietnam could also help Indonesia's rhinos. In 1883, the eruption of Krakatoa triggered a tsunami that leveled the Java forest. What had been a mature forest, containing little for browsing rhinos to eat, rapidly became a secondary forest, which grew full of *Macaranga*, a preferred food plant, along with other pioneer plants the rhinos browsed. Forty years later, enough time has passed that the pioneer plants have given way to mature forest species, which hold less appeal for rhinos. So the fifty or so remaining rhinos in Java may have reached a population ceiling imposed by their food supply. As of 2012, plans were under way to translocate a small number from Ujung Kulon to a second destination on Java.

Vy's efforts to show us an orange-necked partridge, his hospitality, and his delight in being with others who shared his passion for birds and nature offered a different kind of marker of the present in relation to the past than did a bomb crater filled with wallowing elephants. Thirty-three years after departure of the last American soldiers, two American biologists followed their Vietnamese colleague through the forest in search of a rare bird, together trying to find a way to save it from extinction. If there is need to find evidence of hope for the human race, I offer this shred from postwar Vietnam.

Such scraps of optimism are important to cling to, in order to imagine that the Asian nation with the highest concentrations of rare species will manage to preserve many of them. Those cryptic species that remain tucked away in the wettest reaches of the Annamites might be safe from war and its aftermath. But over the seven years that have elapsed since my visit, the string of conservation news from Vietnam has gone from grave concern to tragedy. While writing this chapter, I learned that the last Javan rhino in Vietnam had been killed, shot by a poacher. The official wildlife agency in Hanoi had no comment. Sadly, Vietnam also now has the dubious distinction of being the depot for much of Asia's illegal trade in tiger parts and rhino horn. One reason for the surge in the illegal sale of rhino horn is the emergence there of the false belief that it cures cancer. Beyond extirpating their own rhinos, Vietnam's wildlife criminal gangs are decimating rhino populations in other countries as well. Even the adaptations rare ungulates such as saolas and muntjacs have developed for avoiding leopards, and now humans, will not protect them when commercial hunters carpet the Vietnamese jungles with snares. Snaring for prey is ubiquitous, and in most reserves protection is not enforced. In 2011, for example, three teams of community forest guards in four months collected 7,700 snares from the 220-square-kilometer Saola Nature Reserve in Vietnam's Thua Thien Hue Province.

Do war zones and former war zones always spell decline or disaster for rarities? Sometimes contested areas or the frontlines create no-man's-lands between warring factions where the opposing armies rarely enter. In these buffer zones of relative calm—which also are often heavily mined, as were the borders of Vietnam, Cambodia, and Laos—wildlife can recover. A classic example is the Korean Demilitarized Zone. On this thin strip of heavily mined land, the Korean Peninsula's wildlife, from cranes to deer, has made a comeback. Missing are Amur tigers, but plans are afoot to return them there. An international effort to recover wildlife in former war zones, the Peace Parks Foundation, has as its mission to promote conservation along borders between former combatant nations. There is ample opportunity for this foundation to build a large portfolio of transboundary conservation efforts in the developing world.

At the end of the day, it is fair to ask: How much of what goes on in Vietnam is the legacy of armed conflict, and how much is a disregard for the intrinsic value of nature and rarities? Is Vietnam that much worse than other countries?

It is impossible to test the counterfactual in Indochina—in the absence of decades of war, would rare large mammals still be so endangered and many common species made rare? To do so, one would need to study several replicate Indochinas and hold all other factors constant except for armed conflict in some and long periods of peace in others. Moreover, separating the aftermath of war from combat itself is problematic. What seems irrefutable, however, is that the combined effects of an armed postwar rural populace, lax governance in remote areas regarding hunting and conservation laws, protein shortage, and widespread loss of habitat would likely leave only the small and inedible rarities safe. Even rare endemic land snails could fill a cooking pot. The evidence so far points to Vietnam creating more ghosts. Meanwhile, an adjacent former war zone in Cambodia stands poised to restore past treasures.

~

February 2010. Mondulkiri Protected Forest, eastern Cambodia. We were here to see firsthand another Indochinese landscape attempting to recover from the legacy of decades of armed conflict. Barney Long, a young British biologist working for the WWF, sat in the passenger seat of the air-conditioned Land Cruiser and served as wildlife spotter as we drove along the dusty road from the town of Sen Monorom into the Mondulkiri Protected Forest reserve. At the wheel was Nick Cox, who for the past five years had served as regional coordinator of the WWF's Indochina Dry Forests program. I sat behind them, enjoying the cool, dry air and glad that the thick red road dust coated only the car and not its passengers. How different our journey was from the one Charles Wharton had taken in this same region in the 1950s through the bone-jarring backroads of Cambodia. When his caravan became stuck, the party of sixty members resorted to elephant-back or slogged on foot in search of the wild cattle of backwoods Cambodia.

Suddenly a herd of brown-and-white cow-like creatures galloped across our track—thirteen graceful bantengs, mostly females with calves, and a dominant bull bringing up the rear. By the time we reached the headquarters of the intensive protection area in a village called Mereuch, another herd had crossed our path. "When I first explored this area in 2000," Barney announced, "I saw only two bantengs in twenty-two days." The recovery of rare wild cattle was significant—not only because they are in decline across their range but also because bantengs are an important prey species for tigers. The vast Mondulkiri Protected Forest reserve is one of the few places where tigers are thought to persist in Cambodia.

Ancient Cambodia is best known for Angkor Wat, more than 300 kilometers east of where we were. The civilization of Angkor reached its pinnacle in the thirteenth century, when the city was larger than London and the long-reigning Khmer empire stretched across the landscape we were visiting. If the spectacular temples of Angkor Wat present Cambodia in its finest epoch, the Khmer Rouge regime, Pol Pot, and the killing fields epitomize the nadir of Cambodia's recent past. A country of 4 million people lost half its population during the worst episode of genocide in the past fifty years.

Today Cambodia has the largest wilderness area east of the Mekong River, yet its Eastern Plains landscape is perhaps the best-kept secret among wildlife conservationists in Asia. Covering an area of 18,000 square kilometers, the combined areas of Mondulkiri, Lomphat, Phnom Prich, Snoul, Seima, and Nam Lyr are a remarkable anomaly in this Southeast Asian region characterized by isolated forest fragments. Part of the reason for the integrity of this landscape is that hostilities during the Indochina Wars kept the loggers out. A severe form of malaria common in the area also discouraged development. Further, even though the population of Cambodia has reached 8 million, three-quarters of all Cambodians live in the capital of Phnom Penh. The sparsely populated countryside that once harbored large stretches of unbroken forest has been changing, however. A few years ago—in 2008—Cambodia had one of the highest rates of illegal logging in the world, and it is still high today. According to one observer, forests west of the Mekong were and still are destined for Thailand. Forests east of the Mekong may eventually be cut and sent to Vietnam. Already, the Vietnamese are trucking out as much timber as they can across the porous border. The forests are still extensive in this region, but recovering this lost treasure will require that illegal logging be ended and forest protection improved.

The dry forest, populated by widely spaced trees in the dipterocarp family, the dominant group of trees of Southeast Asia, appeared more like a woodland or savanna. Most trees had shed their leaves during the dry period, adding to the sense of openness. Only along the streambeds did we pass through dense stands of tropical trees and vines. In 1951, Charles Wharton's two-month excursion in the area around Preah Vihear, near where we were, traversed a region filled with large mammals that were rare or extinct elsewhere. Back then, the four species of wild cattle still flourished here, although koupreys were always rare. Wharton filmed six separate groups of koupreys, producing the only existing footage of this species in the wild. He estimated that there were roughly 400 to 500 living west of the Mekong River, 200 to 300 in Lomphat Wildlife Sanctuary, and 50 in the Samrong District of Kratie Province. In 1964, Wharton presented a copy of the film to Prince Norodom Sihanouk, who had a special affection for this rare creature. As a child, Sihanouk had kept a pet kouprey in the Royal Gardens. He subsequently named the kouprey Cambodia's national animal and established Kulen Promtep, Lomphat, and Phnom Prich Wildlife Sanctuaries to protect the last kouprey herds.

Wharton became the champion of efforts to save this species.

After his visit with Sihanouk, he staged another expedition to capture live koupreys for captive breeding as a hedge against extinction in the wild. Although he captured five animals, Wharton ended up with none. Two of them died during handling, and three bolted to freedom. Wharton once said, half joking, that an ancient spell had been cast over the kouprey, shielding it from human efforts to learn about it and save it. In any case, outbreaks of war between the 1960s and the 1980s precluded further kouprey-seeking expeditions. In 1982, a herd was reported near the Thai border, but according to a Cambodian researcher, the effort to find it was called off after a land mine critically injured the group's guide.

New explorers have emerged in the postwar period to walk in Wharton's boots, among them Barney Long. When Barney first visited Mondulkiri, in April 2000, he had the good fortune of working with Lean Kha, a famous resident hunter, former poacher, and veteran of the Cambodian army, which had ousted the Khmer Rouge from this stronghold. Now Kha complained that wildlife was fast disappearing from the forest and suggested that someone pay him to protect the animals rather than shoot them.

On that first survey trip, Barney spent twenty-two days crisscrossing the Mondulkiri forest with Kha and Steven Swan, another British biologist based in Vietnam. Although wildlife was scarce, he sensed potential for recovery. Such a vast, unbroken tract of lowland forest was impossible to find east of the Mekong. If rare cattle and Eld's deer, and along with them the tiger, were to recover anywhere in Cambodia, this would be the place. And recovery seemed possible because tracks of wild cattle were abundant, and tracks of Eld's deer and what could be wild water buffalo were evident, too. The travelers logged their first sighting of the elusive giant ibis; sightings of several species of vultures, populations of which had been decimated across Asia; and slide marks of the rare Siamese crocodile. On the final day of the survey, Barney and his colleagues also saw a tiger track. These promising signs helped to build support for an area wildlife recovery plan. Two years later, the World

Wildlife Fund started working in this Asian wilderness, hiring Kha as a lead ranger to protect the species of Mondulkiri. Reformed poachers often make the best trackers and park guards. They are comfortable with few amenities, and the rough life of the bush is as familiar to them as the morning commute is to urban dwellers. More to the point, they know guns and the mind of the poacher.

A rapid survey in 2010 gave Kha and Barney a chance to reunite and enabled the rest of our group to meet Chana, a Cambodian tiger researcher, and Tom Gray, an ornithologist from the United Kingdom who was transforming himself into a tiger prey specialist. Also with us was Craig Bruce, a wildlife enforcement and protected-area specialist who for the past three years had headed the conservation program in Mondulkiri. Despite the mountains of international funding flowing into Cambodia, little went to wildlife protection: the salaries of the ninety-five-person protection staff in their bright green uniforms, even the uniforms themselves, were courtesy of the WWF and Craig's efforts.

Early the next morning, Kha, Chana, and Barney led our group on a two-hour march across burned stubble to a *trapeang*, or natural water hole, to check on the camera traps. One of Kha, Tom, and Chana's tasks was to map every depression and determine which held permanent sources of water. Permanent water holes are excellent places to set up camera traps to photograph species during the intense dry season. We straggled across the landscape accompanied by two domesticated elephants attached to the patrol staff. Above us, two species of minivets flashed in the morning sunlight. Black-headed orioles chortled their songs and red-breasted parakeets sailed by.

Within a kilometer of the trapeang, a penetrating rolling bugle caught our attention. "Some kind of woodpecker," offered one of the hikers. I knew it was no woodpecker, but I had to click through my memory bank for a few seconds before remembering the unforgettable sound of the sarus crane from my days in Nepal. The cranes had departed before we got to the water hole, but another large silvery-gray bird flew up from a marshy area—a giant ibis! Our early luck was astonishing for even the most jaded in our party.

Tom and Chana walked over to the camera trap and retrieved the footage. They had relied on Kha, who could read game trails invisible to others, to place the cameras for best effect. We were all curious to see the results. With the sun rising in the morning sky, we returned to base.

Back at the station, Tom plugged the memory card into a reader and downloaded the most recent pictures. We gathered around his laptop like eager children waiting to open a new video game. The first images revealed the waterhole regulars: wild boars, barking deer, and civets. There were also carnivores: several kinds of small cats, more civets, and a leopard! No tigers yet, but the team was hopeful, looking forward.

The number and variety of images were impressive. Of particular interest were the pictures of the tiger's prey—gaurs, bantengs, and especially Eld's deer. For decades, first the Vietnamese, then the Khmer Rouge, and then the Cambodian army had used the Mondulkiri area as their larder, living on the meat of the same species tigers prefer to hunt. Therefore, just like the tiger, Eld's deer is endangered. Since the doe usually has only one fawn every two years, it would take some time for this species to recover from overhunting even if given sufficient protection.

As director of the Mondulkiri conservation program, Craig was responsible for protecting the park's wildlife, which primarily meant keeping the animals safe from poachers. Later that day over coffee, Barney recounted how effective that protection had apparently been. When he and Craig went out for a walk in 2008, for example, there they were—bantengs, out in the open during late morning, as calm as could be.

Seeing bantengs lose their fear of grazing in the open during the day was something Barney couldn't have dreamed possible in 2000. He added, "It is amazing what five years of effective law enforcement can do—species can start to recover and even change their behavior." When the human hunters disappear, as they had in this part of Mondulkiri, some formerly common species, such as bantengs, go back to being diurnal. Feeding out in the open without worrying about hunters probably reduced their energy demand and perhaps gave them access to more nutritious grazing areas. The ability to shift back to diurnal grazing might put them on a higher nutritional plane and allow individuals to produce more offspring.

Spotlighting for game at night—as we had in the Cerrado with Carly Vynne (chapter 6)—is always a good way to find rare mammals that move close to roads. We piled into the back of a pickup just before 9:00 p.m., when nocturnal activity is at its peak. The beam of the powerful spotlight arced across the jeep track, seeking reflective eyeshine of whatever might be out there. "There!" whispered someone looking to the left. The light showed a jungle cat intercepted on its evening prowl. The beam was bright enough to illuminate the dark tip of its tail. Along the track we encountered plenty of barking deer but no more Eld's deer: we had all hoped to see a large buck with a dramatic set of antlers. Perhaps once the charred grass layer flushed with new green shoots, the Eld's deer and herds of bantengs and gaurs would be more visible.

The next day, Barney, Nick, and Craig were keen to take us on an afternoon trip on the Srepok River. We stepped into long-prowed boats with extended propeller shafts, the ideal craft for navigating the shallows. As we headed upriver, tree swifts swooped over us, mixed in with needletails, the latter resembling flying cigars. Wreathed hornbills and the more common oriental pied hornbills flapped over the canopy of the riverine forest. Moving between trees were gorgeous red-billed blue magpies, while green malkohas shimmied up the branches. Barney directed us to the best place to see the unusual silver langur monkey. As if on command, a small troop appeared in a fig tree along the river.

The drivers stopped the boats to afford us a better view of the rare primates. Most monkey species love figs, and some hunters attribute monkeys' tasty meat to heavy fig consumption. Consequently, fig-eating monkeys are usually the first mammals to be hunted out of tropical forests. Yet the silver langur seemed locally abundant. I asked, "How did a rare fruit-eating primate prosper in a forest once filled with hunters?" Silver langurs enjoy ingesting the poisonous seeds of a tree, Kha replied. Therefore, local hunters wouldn't waste a bullet on them. The tree species is *Strychnos nux-vomica*, and its seeds contain strychnine. Like the seed-eating saki monkeys of Peru that Sue Palminteri studied, the silver langur makes its flesh toxic to predatory mammals and to humans by eating seeds laced with nasty chemicals.

The presence and future of wildlife in the shadow of war, in Cambodia and beyond, is challenging to predict. Certainly in some buffer areas between warring factions, virtual no-man's-lands laden with mines and dangerous to enter, wildlife hung on. When wars ended, these became source populations to replenish emptied forests. One excellent example is the flooded grasslands of South Sudan, which for much of the year are impenetrable by armies and thus have escaped heavy fighting. The white-eared kob and other large mammals are still abundant there. But in many other war zones in more open areas, where intense fighting has raged, wild animals have been easy targets. Although wars and violence have appeared to decline over the past centuries, according to author Steven Pinker, the diffuse effects on endangered wildlife populations have yet to be realized.

⁓

Vietnam and Cambodia seem to be on different trajectories. Vietnam has abundant conservation plans, but infighting among departments and lack of support from the central government stymie restoration of rarities. Cambodia is more conservation friendly and has more habitat to work with. Cambodia is poised to recover past treasures, but both nations will require more aggressive conservation measures to speed recovery.

Back at the ranger station at Mereuch, we came upon the casing of a cluster bomb dropped by US forces during the Vietnam War. Such sores remain, but the landscape is healing; the soldiers have gone home or, like Kha, have become the backbone of the wildlife protection units. In Cambodia, the ghosts of Indochina can come out of hiding now and regain their lost homeland. The ruins of Angkor Wat feature bas-reliefs of the wild cattle of Cambodia—the banteng, gaur, wild water buffalo, and kouprey. Three of the four remain and have a great future in Mondulkiri, if the conservation world steps up to assist. Cambodian conservationists are taking concrete steps to rebuild the tiger prey base. Once a solid prey base is restored, the government of Thailand, India, or Nepal could, as a goodwill gesture, offer Cambodia a nucleus of adult tigers. Cambodia's Eastern Plains landscape could be the first wild cattle wilderness in the world that is home to a core breeding population of tigers. We can almost see the recovery happening before our eyes. The wildlands sleeping east of Angkor may be about to reawaken.

Chapter 9

Rarity Made Common

*W*E WERE FLYING FROM Calcutta on Druk Air, the national air carrier of Bhutan. When the pilot quietly announced, "Mount Kanchenjunga is now visible from the left side of the plane," the passengers pressed against the windows to photograph the world's third-tallest mountain. Kanchenjunga, whose Tibetan name means "Five Treasures of the Snows," straddles the border of Nepal, the former kingdom of Sikkim (now part of India), and Tibet. The breathtaking panorama of the Himalayas and, just beyond, the beckoning high-plains vista of Tibet kept us glued to the port side. A spirit of serenity wafted like incense through the cabin. In a one-hour flight, the plane had ascended from steamy, sea level Calcutta to the sparkling roof of the world. The next landmark was Mount Jhomolhari. Its gleaming white summit signals entry into Bhutanese airspace and was the cue to return to our seats. The airborne quietude evaporated instantly when the pilot executed a rollercoaster plunge into the tight seam of the Paro Valley. Even among the placid Bhutanese on the plane, this rapid maneuver produces beads of sweat across the brow. Minutes later, the grateful passengers exited the airbus and stood firmly on the tarmac, bathed in the cool, pine-scented breezes of western Bhutan.

This was my second journey to this remote Himalayan kingdom. Most visitors come here to experience the rich cultural heritage or to trek through one of the least disturbed parts of the Himalayas. I came for both purposes, but my main aim on this trip was to inspect a new cultural model in which humans try to live in harmony with nature's rarities rather than contribute to their demise.

Like any other visiting naturalist, I longed to tick off some of the rarities on my list of quest species, in which Bhutan abounds. And then there was Bhutan's fabled conservation program, as progressive, at least on paper, as any in the world, ripe for either an exposé or a paean. But conservation programs go only so far. To the occasional vexation of biologists, the lasting solutions lie in mainstreaming conservation into cultural, economic, and even religious norms. I was here to look at the links between the three: to explore how certain traditions and the norms they embody can protect the habitats of rare species and their populations, enabling them to persist in the modern world. Cleared through passport control and customs—a simple table staffed by one lone, robed official—I was about to find out.

We jumped into a waiting jeep and began the winding hour-long drive from the airport in Paro to the capital, Thimpu. The forested landscape seemed unchanged from my first visit, sixteen years earlier. Two hundred years ago, the Himalayan region was dotted with small kingdoms such as Bhutan, Sikkim, and Gorkha, largely inaccessible to the West and covered, in varying extent, by primeval forest. Nowadays, Himalayan old-growth broad-leaved forests are uniquely preserved in Bhutan, other such examples elsewhere having been chopped down over the past few millennia. Consequently, many animal species that reside in these broad-leaved forests were

Map of the country of Bhutan and the surrounding region

once widespread along the eastern Himalayan chain but are now greatly reduced in range.

Advocates of the Bhutanese way claim that this tiny kingdom of fewer than a million inhabitants offers its neighbors and the rest of the world an inspiring model for living with nature and, by extension, its rarities. Roughly the size of Switzerland, Bhutan was the first nation to establish a permanent fund to finance the long-term protection of its native and rare flora and fauna. More than sixty countries now boast such conservation trust funds, but Bhutan's prototype, capitalized in 1992 with a $1 million seed grant from the World Wildlife Fund, contained a unique feature: the government committed to maintain at least 60 percent of the country under native forest cover, which in turn provides habitat for many rare species.

Along these protected Himalayan slopes, one can still find such rare mammals as red pandas, two species of musk deer, takins, and tigers stalking their prey at timberline, just below where snow leopards roam. The lower reaches of the lush Bhutanese forests are home to golden langurs and beautiful nuthatches, rare species any naturalist longs to see. The golden langur is one of the world's rarest primates, noted for its expressive black face set off by a robe of dense golden fur. These forests are also filled with such splendid birds as brightly colored pheasants and sunbirds and a cacophony of babblers and laughing thrushes, the engaging troubadours of Asian forests. Some say that the rarest primate of all lives here, our apelike cousin the yeti.

The desire to uncover such rareties by several of the most famous nineteenth- and early twentieth-century Western expeditions to the Himalayas to visit Bhutan met with rejection, as have many since. Sir Joseph Dalton Hooker, the famed British botanist, explored neighboring Sikkim in the 1860s but never crossed Bhutan's border, or at least he never admitted to it. William Beebe, an American field biologist, wrote extensively about the magnificent pheasants of the Himalayas, but he had to look for them outside Bhutan. Government policy still discourages foreign-led expeditions today and severely restricts access by individual researchers. Of this they are certain: if there is a rare species that awaits discovery or study, a biology-trained Bhutanese will have the honor.

We arrived in Thimpu at noon under a brilliant blue sky. Traveling with me was my wife, Ute, who had less interest in ibisbills and nuthatches than in Bhutan as a land where Tibetan Buddhism still flourished. As we walked along the streets of the capital, it seemed that, at least outwardly, the dense fabric of Bhutanese cultural and religious life had frayed little in the years since my earlier visit. Whitewashed monasteries gleamed in the sunshine, Tibetan prayer flags fluttered and snapped in the breeze, and men and women walked through town in traditional dress. From the magnificent government centers called *dzongs* to the newest tourist hotels, every piece of architecture still adhered to the national building code: a pagoda-style roof, white walls, and highly decorative painted wood window frames and awnings. Even so, sophisticated Thimpu residents stepped comfortably between two worlds—the modern world of television, cell phones, and Internet access and the traditional one of spinning prayer wheels, oil lamps, and religious festivals.

My first visit, in June 1989, was ill timed, coinciding with torrential monsoon rains. Landslides blocked all road traffic and grounded me in the capital. With no chance to search for rare birds and mammals, I stayed mostly indoors, meeting government officials. One misty afternoon during a break in the weather, I opened my guesthouse window, which looked out on a large meadow. I hoped that a flock of black-necked cranes would circle overhead and magically land in the grass, but it was the wrong season to see Bhutan's most famous endangered species. My fantasy gave way to a different kind of spectacle. Two dozen men clad in brightly colored robes entered the field carrying old-fashioned bows. They broke into two teams, and the archers erected targets about 150 meters opposite each other. The contest began. As arrows flew across the field, each team tried to disrupt the other's concentration with oaths, whistles, and lewd gestures, the men emboldened by swigs from hip flasks. I feared that at least one bowman would be shot dead because the opposing archers stood next to the targets, but their aim was uncanny. From my windowsill seat, the ring of pagodas and banners of prayer flags framed a tableau from the Middle Ages.

By contrast, this dry-season journey of April 2005 promised fine weather in place of monsoons. Himalayan spring had arrived, and Bhutan's spectacular rhododendron forests were bursting into color. The Himalayas are the center of diversity for the genus of "rose trees"; of more than 850 species worldwide, more than 60 are native to Bhutan. Some are dwarf alpine varieties found at timberline, and others, such as those in the Black Mountains, grow thirty meters tall. Some are widespread and abundant, while others have quite localized distributions globally. The narrow ranges of some of these endemic "rhodies" added to Bhutan's long list of rarities. Because Bhutan lies in the eastern zone of the Himalayas, monsoons tend to last longer than in Nepal, and more rain falls. With wetter conditions come more species of plants and, with the incredibly steep topography, more chance for isolation of species. All of these factors greatly increase Bhutan's rarity quotient.

My invitation for this visit came from Kinzang Namgay, head of the WWF's Bhutan program. He had arranged an enticing itinerary, accompanied, as required, by a Bhutanese chaperone. No traveler in Bhutan, or naturalist, goes it alone here. Our journey began with an initial southern excursion, a descent from Thimpu to the edge of Manas National Park. The actual trek would commence at the wintering spot of Bhutan's black-necked cranes, high in the Phobjikha Valley in central Bhutan, and continue through extensive old growth. Aside from the rarities I wished to see and learn about, I was eager for the chance to explore one of the rarest habitats on Earth, extensive tracts of ancient temperate broad-leaved forest.

With an endowment estimated at $30 million in 2012, Bhutan's conservation trust fund pays the annual operating costs of a network of protected areas and biodiversity corridors that together cover almost 35 percent of the country. By government decree, the full Himalayan spectrum—from lowland jungles to snow-covered peaks—receives protection. Far from being stuck in the Middle Ages, in the field of nature conservation Bhutan has bypassed all industrialized nations and emerged in the lead.

The government's decision to ban export logging in its heavily timbered forests is remarkable, since logging, legal and illegal, may be the single greatest threat to the forests of Asia and the persistence of rare species. Logging could be highly lucrative in the short term, but Bhutan's leaders had noticed that in neighboring countries rampant corruption associated with export logging was a cancer, one that could spread rapidly to other sectors of society. In addition, denuded forests and subsequent landslides and siltation would have diminished the country's most profitable export—hydropower sold to India. So, amazingly, they chose to abstain entirely from the destructive industry despite the cash it could bring in.

Bhutan's notable conservation achievements were promoted by a young king who coined the term "gross national happiness" as a novel metric to assess the well-being of his subjects. Protecting a healthy natural environment, conserving the native flora and fauna, and restoring rare species such as cranes and tigers were part of this GNH, along with poverty reduction, literacy, and provision of safe drinking water.

Before we headed off to the mountains, Kinzang had arranged a meeting for me to speak to government officials and to discuss local conservation issues. In most parts of Asia, men tend to dominate government departments, but here the lecture room filled not only with men but also with a surprising number of women. I began my talk by lauding Bhutan's goal to protect all species by protecting large areas of *all* habitats. This approach was similar to the one that John MacKinnon advocated for Vietnam and that underpins conservation strategies in several countries of the region. I suggested that this "representation strategy," as biologists refer to the inclusion of differing habitats, could be complemented effectively by specific programs targeting wide-ranging, migratory, and rare species that might need special measures of protection. In the discussion after my talk, I learned that in the Bhutanese plan, even mythical rare vertebrates rate attention: the main purpose of Sakteng Wildlife Sanctuary was to safeguard the preferred habitat of the yeti.

After the meeting, I was introduced to two well-respected young biologists from the Nature Conservation Division who, to our delight, had volunteered to be our nature guides. Both Nawang Norbu and Sherub had earned master's degrees abroad and were rapidly gaining respect in Bhutan's conservation community. Sherub was one of the country's top ornithologists and, like many Bhutanese, had only one given name (in Tibetan, his name means "wisdom"). Nawang was a talented ecologist who would later head Bhutan's most important forestry institute. The required government-appointed chaperone was Mincha Wangdi, among the kingdom's top environmental educators, an employee of WWF-Bhutan, and the former teacher of Sherub.

Ute had spent the morning exploring one of the large monasteries in Thimpu. She met the attendant lamas and made a donation.

She also purchased Tibetan prayer flags, which saturate the land-scape in a rainbow of blue, green, red, yellow, and white, represent-ing the elements of water, wood, fire, earth, and iron, respectively. The flags are covered in printed prayers and religious symbols, and their display is said to help individuals accumulate merit toward a higher rebirth in their next reincarnation. At another monastery she witnessed the chanting of the monks and spun giant prayer wheels at the entrance to the ceremonial chamber. The massive wheels, nearly four meters tall, contain thousands of written prayers, and the timeless ritual of spinning them in a clockwise motion is in-tended to generate compassion for all living things.

In late afternoon, we drove from Thimpu to a large enclosure for takins, the national animal. Many species figure prominently in Bhutanese mythology, but none more so than the rare takin, an un-gulate of the high mountains now restricted to a small arc around the Eastern Himalayas. According to legend, the takin was created in the fifteenth century when the great Lama Drukpa Kunley, the "Divine Madman," visited Bhutan. The charismatic lama attracted a large following with his unorthodox teachings, startling out-bursts, and magical powers. During his visit he was asked to per-form a miracle. After consuming an entire cow and goat for lunch, he placed the goat's head on the cow's skeleton and waved his arm, and the bizarre takin sprang to life and galloped off to graze.

Our group cautiously approached the tall fence enclosing the captive herd of females, young calves, and a watchful bull. The takin is a stout ungulate about the size of a cow, with a shaggy, gold-hued coat, swollen muzzle, and short, curled horns. Wild takin herds wander the high alpine meadows in summer, feasting on nutritious plants and escaping the leech- and horsefly-infested forests below, keeping a lookout for prowling tigers. Although tigers once occu-pied this entire belt across the Himalayas, now it is only in Bhutan that tigers still venture up to timberline to hunt takins and other montane ungulates. In other countries, tigers are too vulnerable to poachers in such high, remote areas. At the onset of cold weather, wild takins enter the dense forests and disperse. This captive herd lived a pampered, tiger-free life but still seemed quite wild. Our curiosity brought us too close to them, spooking the animals. An adult female barked something in takinese and the whole group rumbled down the hillside.

The next day, the official tour began and we headed for Punakha, the historical capital of Bhutan. As we left the outskirts of Thimpu, we saw a solitary building that declared itself the home of the Karma Insurance Company. I asked Nawang if the standard life insurance policy in Bhutan covers multiple reincarnations. The ensuing string of Buddhist jokes set the tone for the rest of the journey. Merriment seemed to be a favorite pastime in this country, where the citizenry take the concept of happiness very seriously.

National merriment helps offset the mild unease caused by traveling the Bhutanese National Highway, which includes scarcely 200 meters of straight road on the entire route connecting Thimpu in the west with Tashigang in the east. The narrow one-and-a-half-lane highway is carved out of mountainsides and features blind curves, dips, sheer drops, absent guardrails, and long, winding grades that limit vehicles to a snaillike speed.

We continued east, ascending the north–south ridge that separates the conifer forests of western Bhutan from the Himalayan broad-leaved forest in the center and east. Hemlocks grow tall and straight, but their branches droop toward the ground as if weighed down by the heavy growth of old man's beard (the ubiquitous lichen *Usnea*) hanging from the limbs. Himalayan yew, firs, spruces, and pines give off a sweet fragrance. Near the crest of the ridge, conifers defer to stands of hardwoods. Here we stopped to gaze at a natural arboretum of rhododendrons in full glory: pink, magenta, and crimson blossoms brightened lichen-covered tree limbs. The showy white flowers of wild magnolias set off the rhododendron bouquets.

"Let's walk up here for a moment," the soft-spoken Sherub suggested. On a clear day, he said, one can see bearded vultures, or lammergeiers, circling above. "They figure in our custom of sky burial. Our relatives carry the dead to a high promontory, where the body is dismembered and left on a platform for these vultures." The ossifrage, as this raptor is also known, drops bones from great heights so they break, giving the bird access to the marrow. Riding the Himalayan thermals on a three-meter wingspan, the raptor spreads the remains and releases the spirits of devout Buddhists. Unfortunately for us, descending late afternoon mist obscured any chance of seeing one in action, so we continued to Punakha.

At dawn the next day we headed north along a dirt road at the juncture of the Mo and Pho Chu (Male and Female Rivers). Looming over the confluence was the magnificent seventeenth-century Punakha Dzong, a combined fortress and monastery of grand design. This dzong, like others throughout the country, once served as the religious, military, administrative, and social center of the district.

We continued to a quiet spot several kilometers upstream and began scanning the riverbed. Ibisbills are globally rare, but like other rarities we have encountered, such as the rhinoceros in Nepal and the Kirtland's warbler in Michigan, they are often easily encountered in a prime location in the right season, especially when one is accompanied by a skilled guide. "There!" shouted Sherub excitedly. We all jumped out of the vehicle to take a better look through a spotting scope. A pair of ibisbills probed the river's stony margins with their long, curved bills. It is said that ibisbills require clear, fast-flowing water because that is the habitat of the aquatic larvae of the dragonflies, mayflies, and stone flies the birds feed on. One might guess, then, that the cause of the ibisbill's rarity was an absence of pristine rivers. Yet the dirty truth is that a lazy birder can find ibisbills congregating beneath the sewage treatment plants in Thimpu and Paro. So why this species is rare remains a mystery worthy of much further study, including an examination of the effects of hydroelectric projects on these birds' abundance and range. Regardless of their status, ibisbills are captivating to watch. Like the fabled takins, they seemed like another legacy of the

Divine Madman's conjuring: in this case, an ibis's head and bill are mounted on a sandpiper's body that is marked with racing stripes and detailing to create a most striking bird.

Suddenly we were startled by a raucous shriek. Looking up, we saw a huge raptor with a white head and brown body—which could only be the rare, globally threatened Pallas's fish-eagle. A close relative of the bald eagle, it nested in the tall chir pines above the river. While searching for the nest, we spotted two more eagles in the same stretch. Perhaps they were coot hunting. The eagles have been known to land on coots and hold them underwater until they drown before carrying the birds away. But carrion is also a staple food item. The eagles often nest near fishing villages, where they seize scraps and, like the bearded vulture, occasionally sample human corpses. Their low numbers are likely caused by their perch on the top of the Asian food pagoda as large-animal flesh eaters, the jaguars and tigers of the bird world.

An appearance by the white-bellied heron would have completed our morning's rare bird quota, but it was not to be. This species is second only to the goliath heron in size but is much scarcer, although we were told that a pair had been seen using this area. Endemic to the Eastern Himalayas and with perhaps fewer than 250 individuals left, the white-bellied heron is listed as critically endangered in the IUCN Red List. The herons, like the eagles, roost in the tall pines along the river and depend on large, free-flowing rivers, a difficult ecosystem to protect, even in green Bhutan.

Some large herons nest in low vegetation in rookeries, so the white-bellied's propensity for nesting very high in the trees has yet to be explained, although other large herons, such as the great blue and grey, also nest high in trees. At least elsewhere in its range, habitat loss and poaching of adults and eggs are the apparent causes of decline. As a result, the species has been reduced to a few small populations in northeastern India, northern Myanmar, Bangladesh, and eastern Nepal. Extensive hydroelectric development in the region poses a major threat to recovery of these core populations.

As we walked past a bend in the road above the river, a group of Tibetan monks smiled at us as they ambled past and then returned to their chant, fingering their prayer beads. Monks are a common sight in Bhutan, where many young men enter monasteries as children. How different their early lives seemed from ours, I thought. Then I remembered a joke told by a Tibetan friend: "What is the difference between a Buddhist and a non-Buddhist?" The answer: "The non-Buddhist thinks there is one." As the red-robed faithful faded from sight, we turned our attention again to the herons, which remained elusive.

Mincha Wangdi was waiting for us at the guesthouse in Punakha. The next day in the car, we kept up a steady interrogation of our hosts. Sherub and Nawang answered natural history inquiries and the loquacious Mincha handled every other subject, from Buddhism to Bhutanese politics. During a stop for lunch at an outdoor café, the stories rolled off his tongue. My gaze drifted from Mincha to a new sighting, a complement to the national sport of archery. Up the road, a group of young men were hurling what resembled small missiles—carved wooden rockets with nails on the tips—at a target about thirty meters away. This was *khuru*, a Bhutanese version of long-distance darts. As we strolled after lunch along a border of flowering pomegranate trees, the whistles and shouts of the cheerful missileers carried on the breeze.

We made for Trongsa, a tourist attraction because of its beautiful dzong. The highway wound through one of the most exquisite and diverse broad-leaved forests in the temperate regions of the world. Several of the common trees were still leafless but had exploded into flower—the white mimosa-like blossoms covering the *Albizia*, the deep red petals of the coral bean (*Erythrina*), the luscious pink of elephant ear (*Bauhinia*)—all of them accented by the vibrant new foliage of maple, oak, walnut, chestnut, laurel, and, everywhere, the leathery leaves of rhododendron and magnolia. Underneath lay a carpet of valerian, primrose, violet, and club moss dense enough to cushion a sleeping musk deer, or yeti. The forest seemed endless.

Our van snaked around one mountain valley draped in dense forest only to enter the next cove of deep woods. I kept waiting for signs of devastation, a denuded mountainside. Yet nowhere in sight was a mountain village, terraced field, or landslide scar, all inevitable features of the Nepalese and Indian Himalayas.

The mountainous road made us queasy and we had to stop periodically, but each time we were rewarded by the sight of some spectacularly colored birds. First, we admired powder-blue verditer flycatchers, then the shimmering iridescence of orange-bellied leafbirds. Next we saw and heard the blue-throated barbet, and then the most beautiful common bird on any continent—a male scarlet minivet, an elegant treetop inhabitant wearing blushing red and black feathers. But all thoughts of the minivet vanished when a male Mrs. Gould's sunbird landed on the crown of a nearby hemlock. In the brilliant mountain light, its burgundy mantle and back set off its bright yellow rump and belly, highlighted by a long blue tail. The Buddhists say that attachment to beauty is one of the false perceptions humans hold. We, however, suspended Mincha's Buddhist instruction at such moments and wallowed in our attachment.

Over dinner in Trongsa, our conversation ricocheted between natural history and the subject of karma and higher rebirth. I asked Mincha if he would like to be reincarnated as the beautiful scarlet minivet. Mincha paused for a moment and then pointed out how many insects a minivet consumes in its lifetime. "Killing other creatures causes pain in the world. So from a Buddhist perspective, we must say that the minivet is not to be envied." Besides, he related, there are 500 rebirths separating birds and humans, so a bird rebirth would be a big setback from enlightenment.

Back in our room, I reached for a field guide and inadvertently knocked over the rucksack containing our portable library. Across the floor spilled natural history books as well as Ute's copy of *Radical Acceptance: Embracing Your Life with the Heart of a Buddha*, by Tara Brach, a clinical psychologist and respected teacher of Buddhist philosophy and meditation. I dipped into this other type of field guide and soon came upon a passage describing how we live our lives in a mental trance, rarely seeing what is right in front of us. As I continued to read, my evening plan to learn to differentiate laughing thrush species switched to pondering Brach's challenging question to the reader: Is much of one's life spent inside a cocoon of our own making?

The next morning at dawn, we departed from Trongsa and headed south for the village of Zhemgang. As the road descended, we left the cool broad-leaved forest for the lowland warm broad-leaved forest, the richest habitat in Bhutan for birds and perhaps the richest remaining in the entire Himalayas. We quickly saw that this forest was also home to several species of primates. Troops of Assamese macaques crossed the road in front of our car as we scanned the hillsides for langur monkeys.

The golden langur was first reported in 1907. But it was neither photographed nor filmed until the great naturalist E. P. Gee encountered it during several expeditions he made along the Bhutan border in the 1950s. The Zoological Survey of India named the species *Presbytis geei*, or Gee's langur, in his honor (it has since been reclassified as *Trachypithecus geei*).

A common question without an easy answer nagged at me. Why, in the same genus, do we find a ubiquitous species, such as the common or gray langur, which leaps across a wide range of forest canopies of the Indian subcontinent, as well as such rare species as the golden langur, which is confined to a tiny area, and the capped langur, confined to another? Did gray langurs outcompete their golden and capped cousins and limit the latter's range? Or did the common species members, the gray langurs, become isolated and over time evolve into a new species, sometimes with die-off of the parent species in this new locale? Of course, the same question could be posed for any species-rich genus or family, from babblers to barking deer, portrayed in this book. Each large taxonomic grouping in nature seems to have its commoners, adapted to exploit a wide range of habitats, elevations, and tolerance of human presence and its rarities, those family or genus members with a limited range or low abundance. Perhaps for most of these rare members, it was a specialization of habitat or an inability to survive the presence of humans that kept their numbers low. Perhaps the commoners used the available resources in a more efficient fashion. Identifying the causes of rarity in these little-studied species is likely to be a topic of interest for generations of scientists to come.

Here in the lower broad-leaved forests of Bhutan, we were right in the center of the minute range of the magnificent golden lemur, a place where goldens were supposed to be the common species. We kept a close watch on the trees lining the road. Around a bend, we glimpsed the incongruous sight of some fluffy blond bath towels draped over the canopy to dry. Suddenly the towels came to life: black faces, long tails, and the fabled coat of creamy-white and yellow-gold became apparent. Delightedly, we watched the golden langurs jump between trees with wild abandon, tails raised to steady their balance. They appeared to glow, even through binoculars. We exulted as the troop moved across the canopy and down the mountain valley, and then we moved on to a spartan guesthouse in Zhemgang.

The peregrine falcon that flew over the guesthouse as we were leaving Zhemgang the next morning seemed like a good omen. The forest between Zhemgang and our next stop, Tingtibi, lay on the edge of a wildlife corridor connecting Manas, Bhutan's premier lowland national park, and Jigme Singye Wangchuck National Park (formerly known as Black Mountains National Park). The forest was remarkably intact. Young golden langurs frolicked through the treetops while an older animal paused and sat, contemplating an *Albizia* blossom before eating it. The sighting made me rack my brain: where else in the world could one see such a wondrous rarity from the vantage of a paved major thoroughfare?

From the road, we scanned the forest canopy for fruiting fig trees. Ripe figs attract not just monkeys, as noted in the preceding chapter, but other mammals as well. In fact, tropical biologists

*Golden langur (*Trachypithecus geei*) about to feed on an* Albizia *blossom*

estimate that as many as 1,200 species, or roughly 20 percent of all mammals, eat figs for a living. But birds relish them, too, especially large fig eaters such as a bird high on my hope-to-see list for the day, the great hornbill.

On the outskirts of Tingtibi, we parked the truck and made straight for the woods. The trail along the tributary of the great Manas River led through a magnificent unbroken stretch of lowland forest. For the first time in Bhutan, we were sweating in the heat. Resident breeding birds were vocalizing everywhere, as this was the peak of the mating season. Along the trail's border we spied several chestnut-bellied nuthatches trilling away. Nuthatches are a delight to watch feeding, as they are capable of what in humans would be the ultimate party stunt: walking on the ceiling. Nuthatches may forage comfortably all day long while upside down. The riparian forest was brimming with chestnut-bellied nuthatches, right-side-up babblers, and cuckoos, but sadly, no sign of the rare *Sitta formosa*, or beautiful nuthatch. Around noon, the heat of the day quieted the chorus and we split up for the afternoon. Ute and I would drive back with Mincha and his friend in his pickup to spend more time with the langurs and perhaps encounter a beautiful nuthatch at a slightly higher elevation. Sherub and Nawang remained behind to try again along the river. We would all regroup that evening up the mountain in Trongsa.

We didn't drive far before meeting the golden langurs. As we came around a bend, on a rock face bordering the road sat a large troop of mothers with infants, subadults, and an adult male. Dynamite blasting during road construction between Zhemgang and the southern border had exposed a mineral lick where the sharp-featured langurs could restore their electrolyte balance, oblivious to their audience. I was able to sit so close to a troop of one of the rarest monkeys on Earth that I practically could have reached over and brushed their golden fur. Mincha's camera clicked away from the pickup's backseat as the langurs calmly posed for one portrait after another. I could imagine how an ancestral bird similar to the tree creeper could evolve into the beautiful nuthatch, the difference of a brighter blue on top and a warmer orange on the belly. The more highly pigmented body plan of a beautiful nuthatch over that of a more "plain" nuthatch seemed like an easy result of natural selection. But to go from the sooty color of the gray leaf monkey to the brilliant pelage of the golden langur seemed like the alchemy of evolution.

Our driver stopped abruptly at a turnout and pointed to two giant birds sailing like hang gliders over the forest far below. Great hornbills! The birds traversed the entire four-kilometer-long valley, and then moments later a third individual, perhaps an offspring, joined the pair. Through my binoculars, the birds' giant golden-yellow casques, helmetlike protuberances on the bill, and white and black markings stood out against the green blanket of forest. Elsewhere in Asia hornbills are rare, heavily persecuted for their body parts by dealers in traditional medicines. In Bhutan they were still common. The Bhutanese tradition of preserving and respecting life, rather than consuming every last individual of a species, seemed like a natural antidote to the bushmeat trade so pervasive in many countries around the world.

The beautiful nuthatch remained elusive on this leg. But such is often the case in the search for rarities. A pattern we first observed in the Foja Mountains of New Guinea and then saw repeated in the Peruvian Amazon and now in Bhutan is that even in the most intact habitats, many species will always be rare. Birding karma, or even compassion for wild creatures, was not going to trump natural selection and make the beautiful nuthatch any more common or easy to see. Why it has such a spotty distribution over its range is not yet known. Greater clarity on the reasons for such rarity will come when devices such as the TrackTags placed on jaguars become miniaturized to fit smaller rarities, or other new technologies are developed that allow us to gather the data needed to unlock the mysteries of uncommon creatures. This is one of many topics in conservation biology in which theory needs to be bolstered by innovative technologies. Only then will we better understand the more particular conservation needs of many rare species. In the meantime, providing adequate habitat and protection is still the wisest preventive to extinction.

~

Unfortunately, Western explorers of rarity and expatriate conservation biologists in Bhutan are on a fourteen-day visa, just like every other tourist. I could happily have spent another week in lowland Manas National Park, learning about its wildlife and management. But we also wanted to see Bhutan's mountain rarities, so our soon-to-expire travel documents forced us to move to higher ground. Bumthang, at 2,400 meters above sea level and about 66 kilometers east of Trongsa, was the next destination.

The highway climbed once again into the conifer forest as it undulated toward this mountain town famous among Buddhist historians and religious experts. The next day we hiked through a majestic hemlock and juniper forest to a monastery perched on top of a mountain, where the resident lamas scattered grain to attract several species of rare Himalayan pheasants that lived under their protection. For monks and monasteries to offer sanctuary to endangered wildlife is common across the Himalayas. When George Schaller and Peter Matthiessen went on their search for the snow leopard, they made their base camp a remote monastery in the western Nepal Himalayas. There, the head lama had issued a hunting ban. I had seen the results of a similar kind of ban at Tengboche Monastery, below Mount Everest in Nepal, where musk deer and rare pheasants walked without fear of humans because of a lama's decree.

Our next destination was one of the oldest Buddhist temples in the country. Dating from the sixteenth century, this shrine celebrates the arrival of Buddhism to Bhutan via Buddhism's ambassador Guru Rinpoche. The old monastery was empty at the time of our visit, closed for restoration. At a larger, more active monastery nearby, hundreds of maroon-robed monks chanted in the prayer rooms, milled about the courtyards, or basked in the sunshine. Mincha gave us a guided tour and a discourse on monkdom. "Through teachings and meditation," he said, "the lamas seek to cultivate the qualities of gentle kindness, unshakable serenity, and wisdom. Eric, you would benefit by trying it."

Even before Mincha's suggestion, I had been contemplating the interplay of Buddhist teachings and conservation biology. Tara Brach's book had suggested that by offering gentleness and peaceful compassion to all beings, animals as well as humans, the Buddhist philosophy offers a trustworthy route to happiness that is quite different from the route most Westerners follow. Buddhism replaces the pursuit of materialism with a core philosophy of nonattachment and an acceptance of the impermanence of all things. A viewpoint shaped by gentleness and kindness toward all beings— that was how Buddhism could inform conservation biology, it occurred to me. Bhutan's government tried to live that philosophy and had established a unique nation on Earth that is kind to its rare creatures.

I remembered Tim Flannery, the New Guinea mammal expert (chapter 2), and his story predicting the decline of rare mammals there when local animists gave up their old hunting taboos upon their conversion by Christian missionaries. The spread of Buddhist doctrine, as much a gentle philosophy as an organized religion, seemed to have the opposite effect. Here in Bhutan, and in Buddhism in general, the guiding principle was compassion for all living things. Thus a monk spreads grain for rare pheasants. A monastery protects the home range of a musk deer. A head lama declares hunting off-limits in a valley or on an entire mountain range where a snow leopard roamed in search of blue sheep. A fascination with the practice of this philosophy began to rival my interest in local ecology.

∼

We headed west again, retracing our steps toward Thimpu, but then veered off the main road to the evening's destination, the Phobjikha Valley. After a long day of highway driving, we bumped along another twenty kilometers on a dirt track to reach the village's new guesthouse. Although road weary, we perked up after arriving in the wintering area of the black-necked crane, widely revered in Bhutanese culture as a symbol of longevity. The protection of this large migratory crane was what first drew attention to Bhutan among global conservation circles. The marshlands preferred by the cranes could easily have been drained and cultivated, as has happened in many other countries, but instead the Bhuta-

*Black-necked cranes (*Grus nigricollis*) performing a mating dance*

nese government preserved the wetlands as feeding areas for these spectacular giant birds. With just 5,000 to 6,000 individuals of the endangered black-necked species left, the nation's decision to protect this area was a conservation milestone.

Although many long-distance migrants are not considered rare, several of the fifteen crane species have the misfortune of migrating over some of the most hostile terrain on Earth. Such is the case with the western population of the highly endangered Siberian crane, which must traverse areas where guns are prevalent and people are hungry. The birds are harassed and shot during the fall and spring migrations. In contrast, the crane population in Bhutan has increased by 25 percent during the past few years, probably as a result of habitat conservation. Like clockwork, the cranes return each year from their breeding ground on the Tibetan Plateau between October 23 and 26 and stay until late February. Here they are protected by local village committees and the government. The cranes now attract ecotourists, many of whom join the annual festival that welcomes the huge birds back to Bhutan. Schoolchildren dress up in elaborate crane costumes. Although the cranes had already left for their breeding grounds at the time of our visit, we could imagine the vast wetland alive with the birds.

Early the next morning, under threatening clouds, we met our trekking company guide, his staff, and their donkeys. We would spend the next three days on the Gangte path through the Black Mountains; the trail climbs over three passes, weaving through old-growth forests and some of the best upland bird-watching areas in Asia. The route through the forest promised good sightings of satyr tragopans, a rare species of pheasant so named for its head feathers that resemble a satyr's goatlike horns. Everyone admires the most colorful members of the pheasant family; peacocks and ring-necked pheasants belong to this group, as do the crimson fireback and golden pheasants of China. We also hoped to see the Himalayan monal, a huge pheasant cloaked in metallic green, bronze, blue, cinnamon, and purple, with a distinctive wirelike crest of feathers. We listened intently for the whistled "*Kur-leiu*" of the monal and the wailing "*Waah! waah!*" of the tragopan as we began our trek through the forest.

Progress was slow, not so much because of the altitude as because I wanted to stop and look at every bird and plant we passed. Naturalists instinctively know how to practice what Buddhist teachers call the sacred art of pausing. For the naturalist, the best birding sometimes happens when you find a good place to sit and the birds come to you. No more life list, no more fixated desire. Simply wait, hands on binoculars, for something unexpected to happen. For the Western birders on a deadline who all but require ticking so many species on their life list by such-and-such a point in the journey, this philosophy is anathema; the birding trip must be strategized with the cold-blooded efficiency of a US Navy SEAL operation. For the Buddhist, pausing is a way to stop, notice, and then release thoughts and desires that crowd the coop of the hyperactive mind. My allegiance was starting to shift.

Once we had gained the first pass, we dropped into another deeply forested valley. Far from any village, we entered an old-growth hemlock and juniper forest. Here the trees were one and a half meters in diameter and as straight and tall as the masts of great schooners. Underneath was a dense understory of the sweet-smelling cream-colored flowers of *Daphne*, a common ornamental shrub in the United States and in the Himalayas used for paper production. The bright yellow pea flowers of *Piptanthus* lit up the trails. In the forest were rhododendron shrubs as well as trees. We also found azaleas, barberry, wild clematis, and other shrubs whose close relatives appear in gardens back home. As we started up the final pass, we were greeted by a hailstorm. Happily, the storm passed quickly, and the trekking company's guides had hot tea and a bonfire waiting at a camp just below the pass.

The next day's walk to Kokotkha offered another day in the old-growth hemlock forest and yet another dimension to our trip. What had started as a hike became a virtual walking meditation on rarity in nature, but not as one might typically describe a quest for rarity, looking vainly for the last of the last. Instead we were immersed in rarity made common—and enjoyed the uniqueness of being in a place where the species one was observing were rare everywhere else in the Himalayas, because of hunting and habitat loss, but numerous and often rather tame here.

Spotted nutcrackers and spotted laughing thrushes seemed to track our progress through the forest. Above us flew long-tailed minivets, the high-elevation version of the scarlet minivet. We were resting in a forest clearing as the minivets sallied for insects when several large horseflies descended on me. Mincha moved swiftly to dispatch one that had designs on my exposed leg. A startled silence followed. We were unsure of what to make of the swift and sudden execution by a devout Buddhist. "It is only stunned, not dead," he assured us (or perhaps, in denial, assured himself). "Mincha, how *should* a devout Buddhist deal with ectoparasites like horseflies and leeches?" I asked. We were walking through areas that were undoubtedly thick with leeches in the wet season. The resident Buddhist teacher avoided my question.

Over a fabulous dinner of soup, chicken curry, vegetables, and canned fruit and cream, followed by whiskey, we talked about the rare mammals that might be just beyond the tent fly. We had yet to see a musk deer bounding through the oak-rhododendron forests that lay ahead, or a red panda moseying through the bamboo brakes. A Nepalese biologist, Bijaya Kattel, who had studied musk deer in the Everest region of Nepal, discovered the oddest behavior for this deer: they often climbed into the low, spreading branches of trees to feed on lichens. Musk deer in the Himalayas were like tree kangaroos in New Guinea, a species hardly designed for the arboreal niche yet nimble enough to exploit a valuable source of food above ground level. Musk deer males are famous for the scent produced by a gland below their belly. Even the droppings of this primitive deer carry a fragrance.

Tara Brach offers a parable about this species, with some poetic license on wildlife biology:

> A legend from ancient India tells of a musk deer who, one fresh spring day, detected a mysterious and heavenly fragrance in the air. It hinted of peace, beauty and love, and like a whisper beckoned him onward. Compelled to find its source, he set out, determined to search the whole world over. He climbed forbidding and icy mountain peaks, padded through steamy jungles, trekked across endless desert sands. Wherever he went, the scent was there, faint yet always detectable. At the end of his life, exhausted from his relentless search, the deer collapsed. As he fell his horn pierced his belly, and suddenly the air was filled with the heavenly scent. As he lay dying, the musk deer realized that the fragrance had all along been emanating from within himself.

The author commented, "We may spend our lives seeking something that is actually right inside us, and could be found if we would only stop and deepen our attention."

After more whiskey, it was time to air out the yeti stories. No rare Himalayan species has garnered more ink, launched such fruitless expeditions, or attracted more cranks than this rarity, the "abominable snowman." Search parties have scoured parts of the Himalayas and brought back traces of fur claimed to be from a yeti, but these typically turn out to be from a Himalayan tahr—a relative of the Rocky Mountain goat—a Himalayan brown or black bear, a langur, or a macaque. The absence of scientific proof does little to dissuade the locals, who are convinced that the yeti, known here as the *migoi*, is real. In fact, the Bhutanese add a novel feature to its repertoire—the ability to become invisible when necessary. Such revelations only add fuel to the beliefs of diehard cryptozoologists (literally, those who study hidden creatures) who remain convinced it is only a matter of time before hard evidence convinces a skeptical world.

The next day, after crossing another mountain pass, we were at last entering old-growth oak and rhododendron forest. At the next rise, I walked clockwise around a stone monument, a religious shrine, and leaned against a boulder. As soon as I put my pack down, a cascade of cuckoo songs tumbled down the mountainside. The large hawk cuckoo is an incessant singer. Repeatedly, I heard its definitive self-diagnosis: "*Brain fe-ver, brain fe-ver, brain fe-ver,*" consoled by a nearby oriental cuckoo's soft "*Ho-ho-ho-ho*"; offered treatment by the Indian cuckoo's "*One-more-bottle, one-more-bottle*"; and dismissed by the psychoanalytic Eurasian species with the classic rejoinder "*Cuck-koo, cuck-koo.*" I tried to call in the oriental but instead stirred up a juvenile yellow-billed blue magpie.

The oaks were massive, draped in lichens. Rhododendrons colored the scene—the giant *Rhododendron arboreum*, with its bright red flowers and the most beautiful variety I had ever seen; the species *R. hodgsonii*, now all around me, its trunk and limbs festooned with long peeling strips of purplish bark. The rhododendrons had peaked, and the trail lay strewn with fallen flowers. William Beebe's quote "To be a Naturalist is better than to be a King" floated through my mind.

To pause in this old-growth oak-rhododendron forest, to be in this moment, seemed to offer a perfect marriage of Buddhist practice and scientific curiosity. There is something wondrous about walking among living organisms many hundreds of years older than you. Deep groves of old growth are globally rare and can induce a state of rapture in those open to the experience, whether in the redwoods or sequoias of California, the hill dipterocarp forests of Sarawak, the mountain ash stands of Australia, the venerable hemlock-cedar forests of Vancouver Island, or the primeval koa stands of Hawaii. And rapture mixes with tranquility, an inner quietude that envelops the soul of every nature lover who enters a valley of ancient trees. Standing on a petal-strewn path in this rare forest offered a taste of what practiced Buddhists must feel when deep in meditation. A phrase echoed in my head, a phrase that naturalists in nature know but sometimes fail to name. Like the repetitive refrains of the cuckoos, it was a welcoming, unshakable song: "*Serenity, serenity.*"

My reverie ended with the sudden arrival and rapid departure of two Americans striding down the trail. Their Bhutanese guide, who was puffing along behind them, muttered that the Yanks had decided to cover in one day's walk what our party would do in three. So much for the sacred art of pausing. The racers failed even to slow down to marvel at the serenade of cuckoos, or the intensity of the purple-barked rhododendrons, or the calming effect of standing in an old-growth forest. Perhaps the Buddhists are right in observing the nature of impermanence underlying everything of this Earth, but I wanted to burn into every neuron what it was like to stand in such a magnificent forest. I wanted an image that would last decades, remaining with me when I was too old to climb this ridge again. My sense of moral superiority evaporated when Mincha struggled up to where I was and asked if I had seen the American hikers. "You mean the ones who raced through with blinders on?" "Oh yes," he replied. "They had four satyr tragopans cross the trail in front of them. You must have just missed it."

Now the trail began descending rapidly, and before long we would be out of the altitudinal range of the evasive tragopans. Up ahead, a shaggy gray-coated mammal bounded across the trail. A juvenile or yearling yeti? The upright, wagging tail ultimately revealed a village dog that came up to greet us. We walked along together until we caught up with the cook, who had a hot lunch waiting. I sat on a rotting log with hemlock and rhododendron seedlings, sedges, and ferns growing out of it. Finding the dog hungry, I shared my potato pea curry with the mutt. Later I came across an appropriate Bhutanese proverb, "If merit is to be earned, be good and kind to dogs."

There was much to like about this country and its customs, including kindness to dogs. In other Asian countries, dogs may wander half starving and mange afflicted; the Bhutanese, by contrast, are typically generous with food and care. Dogs are considered a high rebirth, next to humans in the chain, bumping the apes back down the list despite all the DNA evidence. The Bhutanese believe that dogs have intervened on behalf of humans when the gods were angry with them. Dogs are also said to be helpful in the afterlife: if a human soul is lost in the darkness of the hereafter, dogs show the way with a light glowing on their tails. Just above the base of the mountain, my new companion returned my favor. He darted under some brush and scared up a Kalij pheasant, our first of the journey. Moments later, we were at the edge of a village and a few minutes from the waiting van.

～

Our departure from Bhutan was going to generate a painful withdrawal. For two weeks, life had slowed to the speed at which I believe we are meant to live. The fresh air, the altitude, the powerful influence of the Buddhist culture seemed to awaken each of us from the trance, the self-made cocoon we lived in back home. If only for a short while, we had escaped to a different place, a geography where a pair of hiking shoes, binoculars, and a cup of hot tea seemed like enough.

As we packed away our gear and fond memories, a question raised by Bhutan's critics popped up: What did a small, remote, sparsely populated and still untouched country have to teach the rest of the world about conservation, reverence for rare species, or anything? Is it really an outlier among nations?

But the critics ask the wrong question. It is not the size of the country or its intactness per se, but the philosophy that guides it, that is important. The Bhutanese have taken the principles of modern conservation biology and woven them into the Buddhist dharma to chart a different course for their nation. So perhaps a better question is: What solution does a devoutly Buddhist culture offer for the conservation crisis? The answer: The global conservation crisis is ultimately a spiritual crisis in disguise. And what we lack in abundance is compassion for the millions of other species with which we share the planet, something that comes as naturally to the Bhutanese as breathing. Perhaps that is the country's most essential export to the rest of us who are trying to come to grips with the conservation of rarities.

Even in a nation where the majority of civic and religious leaders and its populace express compassion for all living things, the record is not perfect, of course. Overzealous government officials can make regrettable decisions. In 2012, several years after we completed this journey, someone in the government granted permission to "improve" the trail we had hiked on and turn it into a road with a power line. Some of the massive oaks lay strewn along the wayside, casualties in the name of progress. Fortunately, much of the old-growth forest remained intact adjacent to it, but this story illustrates how, in the absence of constant vigilance, a few individuals can make decisions against the best interests of a nation.

~

Our cultural evolution as a species is in its adolescence. But evolution never stops, and perhaps ahead of us is a prominent marker in our own development: the point when we truly value nature's diversity, a metric noted by conserving rare wildlife. And as with the dying musk deer, the answer to our dilemma of how to take that next step was right in front of me, right in front of us, all along. Developing our gift for compassion is a critical contribution to the persistence of rarities.

Compassion alone, of course, is insufficient. It didn't work for flightless birds against invading rats in Hawaii, nor will it save many other of nature's rarities. Incentives, economic or otherwise, for conservation, superb science, and improved governance for everything from a climate change treaty to enforcement of antipoaching laws are also necessary parts of the solution. The fate of rarities is not only in the hands of impoverished villagers but also in the hands of those in political palaces and the boardrooms of multinational corporations who could take seriously the conservation of rarity and act on its behalf with far-reaching effect. The combined actions of Big Agriculture, for example, have far greater consequences for the persistence—or extinction—of rarities than do the effects of indigenous groups scattered throughout the tropics.

The challenge ahead for us in preserving rarities is to link the science-based approach that focuses on populations rather than individuals and the animal-welfare philosophy that gives ethical value to individuals and their well-being. There is ample evidence of reason to hope for such a grand merger of science-based thinking and compassionate connection to wildlife. We see the response of compassion in laws preventing animal cruelty and in the growth of rescue shelters for dogs, cats, and wild animals. The combined scientific and compassionate response is also taking root. A global tiger summit, the International Forum on Tiger Conservation, staged in November 2010 and attended by heads of state of the tiger range countries—the first ever such forum for a wild species—may give this rare top carnivore a second chance through its commitment to double the wild tiger population by 2022. In 2012, new legislation was passed in several countries to stop the finning of sharks, another top predator that has been made rare by senseless slaughter, and whose decline has altered the regulation of marine systems.

Some critics hold that wildlife conservation will be an unattainable luxury for the poorest countries until their citizenry can climb out of poverty. Besides Bhutan, the examples of two other economically poor countries counter such an assertion. Nepal, as we've seen, is now doing a better job of protecting its rare endangered vertebrates than are most countries. Perhaps most successful of all is Namibia, another country characterized by extreme rural poverty yet soon to have almost half its land area covered by communal conservancies and national parks protecting many rarities of that nation's arid lands, from succulent plants in the Namib-Karoo region to free-ranging black rhinos. If three of the poorest nations on Earth have some of the best track records for conserving rarities, other factors must be at work.

A marriage of science, political will, and compassion in rich countries as well as poor could embrace not only empathy for animals but also our ability to conduct and understand science and to appreciate fully the wonderful beauty of rare animals and old-growth forests, the complexity of life, the immensity of the universe. As we enter the Anthropocene epoch, we must hope that humans in the early twenty-first century will finally be able to reach accommodation with uncommon nature and—dare we hope—a celebration of rarities.

Annotated Bibliography

For those wishing to delve deeper into the subject of rarity, I have selected a number of books, articles, and scientific papers that I drew upon in my own research or that expand greatly beyond the species I have covered. Under each chapter title, I have tried to identify references that might appeal to a broad range of readers, professionals, and students.

Chapter 1. The Uncommon Menagerie
Dinerstein, Eric. *The Return of the Unicorns: The Natural History and Conservation of the Greater One-Horned Rhinoceros.* New York: Columbia University Press, 2003.
A monograph covering most aspects of the biology and conservation of a megafauna species, including interactions with its environment.

Gaston, Kevin J. *Rarity.* London: Chapman and Hall, 1994.
An excellent introduction to the biology of rarity by perhaps the world's foremost authority on the subject.

———. *The Structure and Dynamics of Geographic Ranges.* Oxford: Oxford University Press, 2003.
This work provides an excellent background to the science of species' geographic distributions and abundances.

Kunin, William E., and Kevin J. Gaston, eds. *The Biology of Rarity: Causes and Consequences of Rare-Common Differences.* London: Chapman and Hall, 1997.
These essays, by some of the world's leading authorities on rarity, are technical but thought provoking and challenge many assumptions about rarity. A great place to start for any scientist or student of the field.

Ricketts, Taylor H., Eric Dinerstein, Tim Boucher, Thomas M. Brooks, Stuart H. M. Butchart, Michael Hoffmann, John F. Lamoreux, et al. "Pinpointing and Preventing Imminent Extinctions." *Proceedings of the National Academy of Sciences of the United States of America* 102, no. 51 (2005): 18497–501.

The scientific paper that underpins the work of the Alliance for Zero Extinction. The organization's website, http://www.zeroextinction.org, contains useful information about rare vertebrates and some plants (conifers).

Chapter 2. The Gift of Isolation

Beehler, Bruce M. "The Lost World." *Living Bird* 25, no. 4 (2006): 15–24. Conservation International's Rapid Assessment Program expedition to the Foja Mountains of Indonesia New Guinea.

————. *Lost Worlds: Adventures in the Tropical Rainforest.* New Haven, CT: Yale University Press, 2008.
A naturalist's travels in tropical countries around the world for research and nature conservation, with a special focus on New Guinea.

Diamond, Jared M. "Distributional Ecology of New Guinea Birds: Recent Ecological and Biogeographical Theories Can Be Tested on the Bird Communities of New Guinea." *Science* 179, no. 4075 (1973): 759–69.
A magisterial overview of the ecology and distribution of the rich bird fauna of the island of New Guinea.

Frith, Clifford B., and Bruce M. Beehler. "The Birds of Paradise." *Scientific American* 261 (December 1989): 117–23.
A review of the ecological underpinnings of the polygamous mating systems of birds of paradise.

————. *The Birds of Paradise: Paradisaeidae.* New York: Oxford University Press, 1998.
A monographic treatment of the biology of the birds of paradise.

Frith, Clifford B., and Dawn W. Frith. *The Bowerbirds: Ptilonorhynchidae.* New York: Oxford University Press, 2004.
A monographic treatment of the biology of bowerbirds.

Gressitt, J. Linsley, ed. *Biogeography and Ecology of New Guinea.* 2 vols. The Hague: W. Junk, 1982.
A multiauthored treatment of the ecology and biogeography of the island of New Guinea.

Marshall, Andrew J., and Bruce M. Beehler, eds. *The Ecology of Papua.* 2 vols. Singapore: Periplus, 2007.
Fifty-two chapters on the natural history of western New Guinea, contributed by more than sixty expert authors.

Wallace, Alfred Russel. *The Malay Archipelago, the Land of the Orang-utan and the Bird of Paradise: A Narrative of Travel, with Studies of Man and Nature.* New York: Dover, 1962.
One of the great works of natural history travel and study by one of the cofounders of the field of evolution.

Chapter 3. A Jaguar on the Beach

Asner, Gregory P., George V. N. Powell, Joseph Mascaro, David E. Knapp, John K. Clark, James Jacobson, Ty Kennedy-Bowdoin, et al. 2010. "High-Resolution Forest Carbon Stocks and Emissions in the Amazon." *Proceedings of the National Academy of Sciences of the United States of America* 107, no. 38 (2010): 16738–42.
The first published account of an effort to map forest carbon across a large tropical landscape.

Colinvaux, Paul A. *Why Big Fierce Animals Are Rare: An Ecologist's Perspective.* Princeton, NJ: Princeton University Press, 1978.
An excellent explanation of the costs of being an apex predator.

Dinerstein, Eric, Keshav Varma, Eric Wikramanayake, George Powell, Susan Lumpkin, Robin Naidoo, Mike Korchinsky, et al. "Linking Ecosystem Services, Conservation, and Local Livelihoods through a Wildlife Premium Mechanism." *Conservation Biology* (in press).
The wildlife premium mechanism is a new performance-based approach to link conservation of carbon held in rain forests and recovery of endangered wildlife such as jaguars, tigers, and elephants.

Estes, James A., John Terborgh, Justin S. Brashares, Mary E. Power, Joel Berger, William J. Bond, Stephen R. Carpenter, et al. "Trophic Downgrading of Planet Earth." *Science* 333, no. 6040 (2011): 301–6.
An important review paper that identifies an extensive suite of "ecological surprises"—unanticipated impacts on ecosystems ranging from tundra to coral reefs and ecological processes ranging from wildfire to disease—stemming from the loss of apex consumers.

Forsyth, Adrian, and Ken Miyata. *Tropical Nature: Life and Death in the Rain Forests of Central and South America*. New York: Simon and Schuster, 1987.
One of the classic natural history accounts of rain forests.

Foster, Robin B. "The Floristic Composition of the Rio Manu Floodplain Forest." In *Four Neotropical Rainforests*, edited by Alwyn H. Gentry, 99–111. New Haven, CT: Yale University Press, 1990.
A description of the vegetation in one of richest rain forests on Earth.

Nuñez-Iturri, Gabriela, Ola Olsson, and Henry F. Howe. 2008. "Hunting Reduces Recruitment of Primate-Dispersed Trees in Amazonian Peru." *Biological Conservation* 141, no. 6 (2008): 1536–46.
The negative impacts of hunting of primates on the regeneration of trees is the subject here, and how the loss of seed dispersers leads to changes in tree species composition that are independent of logging or other human activities in the forest.

Palminteri, Suzanne, George Powell, Whaldener Endo, Chris Kirkby, Douglas Yu, and Carlos A. Peres. "Usefulness of Species Range Polygons for Predicting Local Primate Occurrences in Southeastern Peru." *American Journal of Primatology* 73, no. 1 (2011): 53–61.
This paper shows how range polygons, which are generated at the scale of species' geographic ranges, can be used for conservation planning at the project scale (usually a single landscape).

Sanderson, Eric W., Kent H. Redford, Cheryl-Lesley B. Chetkiewicz, Rodrigo A. Medellín, Alan R. Rabinowitz, John G. Robinson, and Andrew B. Taber. "Planning to Save a Species: The Jaguar as a Model." *Conservation Biology* 16, no. 1 (2002): 58–72.
The first expert-driven effort to gather current (1999) scientific knowledge about jaguars—their known geographic range, conservation status, and chances for survival—at the continental scale.

Terborgh, John. *Five New World Primates: A Study in Comparative Ecology*. Princeton, NJ: Princeton University Press, 1983.
The ecology of five primates in Manú National Park, Madre de Dios, Peru, is featured in this book. The author also explains the underlying environmental processes that influence these species in their movement, including feeding, predator avoidance, and other behaviors.

Terborgh, John, Lawrence Lopez, Percy Nuñez, Madhu Rao, Ghazala Shahabuddin, Gabriela Oriheula, Mailen Riveros, et al. "Ecological Meltdown in Predator-Free Forest Fragments." *Science* 294, no. 5548 (2001): 1923–26.
A classic paper in ecology that documents the changes to an ecological system stemming from the exclusion of predators on forested islands created by a dam in Venezuela. The resulting herbivore population explosion led to lower recruitment of favored tree species and subsequent changes in plant composition, which in turn decreased the amount of food available to the herbivores, leading to a loss of diversity at several trophic levels.

Tobler, Mathias W., Sarnia E. Carrillo-Percastegui, Renata Leite Pitman, Rosa E. Mares, and Glen E. Powell. "An Evaluation of Camera Traps for Inventorying Large- and Medium-Sized Terrestrial Rainforest Mammals." *Animal Conservation* 11, no. 3 (2008): 169–78.
A study in the Los Amigos River basin evaluates the efficiency of camera trapping for inventorying larger rain forest mammals and estimating species richness at a given research site.

Ziegler, Christian, and Egbert Giles Leigh Jr. *A Magic Web: The Forest of Barro Colorado Island.* Oxford: Oxford University Press, 2002.
The best introduction available to tropical rain forests, accompanied by stunning photographs.

Chapter 4. The Firebird Suite
Askins, Robert A. *Restoring North America's Birds: Lessons from Landscape Ecology.* New Haven, CT: Yale University Press, 2004.
Overview of the characteristics and nature of management required for the Kirtland's warbler on its breeding grounds.

Deloria-Sheffield, Christie M., Kelly F. Millenbah, Carol I. Bocetti, Paul W. Sykes Jr., and Cameron B. Kepler. "Kirtland's Warbler Diet as Determined through Fecal Analysis." *Wilson Bulletin* 113, no. 4 (2001): 384–87.
A useful guide to what this endangered songbird eats.

Mayfield, Harold. *The Kirtland's Warbler.* Bloomfield Hills, MI: Cranbrook Institute of Science, 1960.
Good natural history of this species.

Probst, John R., and Jerry Weinrich. "Relating Kirtland's Warbler Population to Changing Landscape Composition and Structure." *Landscape Ecology* 8, no. 4 (1993): 257–71.
The relationship between breeding success in Kirtland's warblers and landscape characteristics, especially the birds' reliance on young jack pine stands, is made clear in this paper.

Rabinowitz, Deborah, Sara Cairns, and Theresa Dillon. "Seven Forms of Rarity and Their Frequency in the Flora of the British Isles." In *Conservation Biology: The Science of Scarcity and Diversity*, edited by Michael E. Soulé, 182–204. Sunderland, MA: Sinauer Associates, 1986.
One of the foundational papers on rarity in nature by a leading thinker. Rabinowitz was the first to combine a species range, population number, and habitat specificity to define various forms of rarity in nature, tested against the flora of the British Isles.

Walkinshaw, Lawrence H. *Kirtland's Warbler: The Natural History of an Endangered Species*. Bloomfield Hills, MI: Cranbrook Institute of Science, 1983.
Another classic natural history of this species.

Wunderle, Joseph M., Jr., Dave Currie, Eileen H. Helmer, David N. Ewert, Jennifer D. White, Thomas S. Ruzycki, Bernard Parresol, and Charles Kwit. "Kirtland's Warblers in Anthropogenically Disturbed Early-Successional Habitats on Eleuthera, the Bahamas." *Condor* 112, no. 1 (2010): 123–37.
Provides an overview of the Kirtland warbler's winter habitat and factors that affect the production of the winter habitat and its conservation.

Chapter 5. There in the Elephant Grass
Dinerstein, Eric. "Effects of *Rhinoceros unicornis* on Riverine Forest Structure in Lowland Nepal." *Ecology* 73, no. 2 (1992): 701–4.
How forests with rhinos look different from forests without them.

———. "Family Rhinocerotidae (Rhinoceroses)." In *Handbook of the Mammals of the World*, vol. 2, *Hoofed Mammals*, edited by Don E. Wilson and Russell A. Mittermeier, 144–81. Barcelona: Lynx Edicions, 2011.
A recent review of the ecology and conservation of the five extant rhinoceros species.

————. "Seed Dispersal by Greater One-Horned Rhinoceros (*Rhinoceros unicornis*) and the Flora of *Rhinoceros* Latrines." *Mammalia* 55, no. 3 (1991): 355–62.
Rhinoceroses as major fruit eaters and landscape engineers.

Dinerstein, Eric, and Chris M. Wemmer. "Fruits *Rhinoceros* Eat: Dispersal of *Trewia nudiflora* (Euphorbiaceae) in lowland Nepal." *Ecology* 69, no. 6 (1988): 1768–74.
An account of megafaunal fruits and megafauna in Asia.

Owen-Smith, R. Norman. *Megaherbivores: The Influence of Very Large Body Size on Ecology.* Cambridge: Cambridge University Press, 1992.
A classic account of the advantages and disadvantages of body mass in large terrestrial mammals.

Peters, Robert H. *The Ecological Implications of Body Size.* Cambridge: Cambridge University Press, 1983.
A stimulating book on a topic central to the ecology of pachyderms.

Chapter 6. Scent of an Anteater
Brannstrom, Christian, Wendy Jepson, Anthony M. Filippi, Daniel Redo, Zengwang Xu, and Srinivasan Ganesh. 2008. "Land Change in the Brazilian Savanna (Cerrado), 1986–2002: Comparative Analysis and Implications for Land-Use Policy." *Land Use Policy* 25, no. 4 (2008): 579–95.
This paper discusses the implications of the Forest Code for past and projected land-use patterns in the Cerrado.

Cremaq, Piauí. "Brazilian Agriculture: The Miracle of the Cerrado." *Economist,* August 26, 2010, http://www.economist.com/node/16886442.
This article provides a perspective on how the Cerrado became the new midwestern United States and allowed Brazil to emerge as a global agricultural superpower.

Emmons, Louise H., ed. *The Maned Wolves of Noel Kempff Mercado National Park.* Smithsonian Contributions to Zoology, no. 639. Washington, DC: Smithsonian Institution Scholarly Press, 2012.
Ecology of maned wolves in Bolivia.

Klink, Carlos A., and Ricardo B. Machado. "Conservation of the
    Brazilian Cerrado." *Conservation Biology* 19, no. 3 (2005): 707–13.
A fine overview of the biological importance of the Cerrado and the
major threats to its conservation.

Oliviera, Paulo S., and Robert J. Marquis. *The Cerrados of Brazil: Ecology
    and Natural History of a Neotropical Savanna.* New York: Columbia
    University Press, 2002.
An important book on the ecology of the Cerrado, including chapters
on diverse topics such as land use, plant communities, the role of fire in
the system, and animal community diversity and natural history.

Silva, José Maria Cardoso da, and John M. Bates. "Biogeographic
    Patterns and Conservation in the South American Cerrado: A
    Tropical Savanna Hotspot." *BioScience* 52, no. 3 (2002): 225–34.
This paper introduces the Cerrado's savannas in a global context and describes the main biogeographic patterns. It also introduces the principal
vegetation types and highlights the origin and evolution of species diversity and endemism in the region.

Silveira, Leandro, Anah Tereza de Almeida Jácomo, Mariana Malzoni
    Furtado, Natália Mundim Tôrres, Rahel Sollmann, and Carly
    Vynne. "Ecology of the Giant Armadillo (*Priodontes maximus*) in
    the Grasslands of Central Brazil." *Edentata*, nos. 8–10 (2009): 25–34.
One of the few scientific discussions of the giant armadillo. The most
extensive study to date, recently conducted in Emas National Park, is
reported in this paper.

Tollefson, Jeff. "Brazil Revisits Forest Code." *Nature* 476 (August 17,
    2011): 259–60.
In 2011, Brazil revisited its 1965 Forest Code, which has been a corner-
stone in the country's environmental protection efforts. The current law
being revisited requires that landowners in the Cerrado must maintain
20–35 percent of their land (depending on the state) in a natural state
and that those who had cleared illegally must reforest to that level.

Vynne, Carly, Jonah L. Keim, Ricardo B. Machado, Jader Marinho-
    Filho, Leandro Silveira, Martha J. Groom, and Samuel K. Wasser.
    "Resource Selection and Its Implications for Wide-Ranging
    Mammals of the Brazilian Cerrado." *PLoS ONE* 6, no. 12 (2011).

Results of a field study of the landscape features selected by the giant armadillo, giant anteater, puma, jaguar, and maned wolf in and around a nature reserve in the Brazilian Cerrado. Conservation of these five wide-ranging species will require prioritizing the landscape features and composition requirements identified in this paper and ensuring that these features are maintained, protected, and restored.

Vynne, Carly, John R. Skalski, Ricardo B. Machado, Martha J. Groom, Anah T. A. Jácomo, Jader Marinho-Filho, Mario B. Ramos Neto, et al. "Effectiveness of Scat-Detection Dogs in Determining Species Presence in a Tropical Savanna Landscape." *Conservation Biology* 25, no. 1 (2011): 154–62.
All about the effectiveness of using scat detection dogs to study rare wide-ranging mammals in the Brazilian Cerrado. The distributions of giant armadillos, giant anteaters, pumas, jaguars, and maned wolves in and around Emas National Park, in the Brazilian Cerrado, are also reported.

Chapter 7. Invasion and Resistance

Hargreaves, Dorothy, and Bob Hargreaves. *Tropical Trees of Hawaii*. Honolulu: Island Heritage, 1964.
A nice pictorial guide to Hawaii's native and exotic trees.

Hawaii Audubon Society. *Hawaii's Birds*. Honolulu: Hawaii Audubon Society, 1997.
Excellent field guide.

Pratt, Thane K., Carter T. Atkinson, Paul C. Banko, James D. Jacobi, and Bethany L. Woodworth, eds. *Conservation Biology of Hawaiian Forest Birds: Implications for Island Avifauna*. New Haven, CT: Yale University Press, 2009.
A veritable encyclopedia of Hawaiian forest birds, full of useful information about their history and ecology and efforts to recover extant species.

Quammen, David. *The Song of the Dodo: Island Biogeography in an Age of Extinctions*. New York: Simon and Schuster, 1997.
A highly readable popular account of the biology of true islands and mainland islands and the effects of fragmentation on extinction of populations.

Soehren, Rick. *The Birdwatcher's Guide to Hawaii*. Honolulu: University of Hawaii Press, 1996.
Excellent compact field guide.

Wagner, Warren L., and Vicki A. Funk, eds. *Hawaiian Biogeography: Evolution on a Hot Spot Archipelago*. Washington, DC: Smithsonian Institution Press, 1995.
Covers the radiations of Hawaii's most interesting groups and is a great source of evolutionary insights.

Weiner, Jonathan. *The Beak of the Finch: A Story of Evolution in Our Time*. New York: Alfred A. Knopf, 1994.
Excellent account of adaptive radiation and speciation among Galápagos finches.

Wilson, Edward O. "The Nature of the Taxon Cycle in the Melanesian Ant Fauna." *American Naturalist* 95, no. 882 (1961): 169–93.
The introduction of the concept of the taxon cycle as applied to a well-studied and widespread group—ants.

Chapter 8. Ghosts of Indochina
Baltzer, Michael C., Thi Dao Nguyen, and Robert G. Shore. *Towards a Vision for Biodiversity Conservation in the Forests of the Lower Mekong Ecoregion Complex*. Hanoi: WWF-Indochina; Washington, DC: WWF-US; Gland, Switzerland: WWF International, 2001.
This seminal work maps the biogeography of the Lower Mekong Dry Forests Ecoregion and prioritizes landscapes for conservation action. It has been the basis for a range of conservation programs and regional planning documents.

Critical Ecosystem Partnership Fund. *Ecosystem Profile: Indo-Burma Biodiversity Hotspot; Indochina Region*. Final version. Arlington, VA: Conservation International, Critical Ecosystem Partnership Fund, May 2007. http://www.cepf.net/Documents/final.indoburma_indochina.ep.pdf.
An up-to-date snapshot of conservation priorities in the Indo-Burma region, covering both landscapes and species.

Duckworth, J. William, and S. Blair Hedges. *Tracking Tigers: A Review of the Status of Tiger, Asian Elephant, Gaur, and Banteng in Vietnam, Lao[s], Cambodia, and Yunnan (China), with Recommendations for*

Future Conservation Action. Hanoi: WWF Indochina Programme, 1998.
A startling older account of the rarity of very large mammals in the Indochina region; the situation is even bleaker now.

Duckworth, J. William, Richard E. Salter, and Khamkhoun Khounboline, eds. *Wildlife in Lao PDR: 1999 Status Report*. Vientiane: IUCN— World Conservation Union, Wildlife Conservation Society, and Centre for Protected Areas and Watershed Management.
The definitive book on the wildlife of Lao PDR, its distribution, and its status. Although now an older text, this remains the go-to reference on wildlife in Lao PDR.

Hardcastle, James, Steph Cox, Thi Dao Nguyen, and Andrew Grieser Johns, eds. *Rediscovering the Saola: Proceedings of "Rediscovering the Saola—A Status Review and Conservation Planning Workshop."* Hanoi: WWF Indochina Programme, 2005.
An overview of the status and ecology of Indochina's flagship species, the saola. This document outlines a regional plan of action to recover the species before its extinction.

IUCN-SSC Asian Wild Cattle Specialist Group. *Regional Conservation Strategy for Wild Cattle and Buffaloes in South-east Asia*. Gland, Switzerland: International Union for Conservation of Nature, Species Survival Commission, 2010.
A clear plan of conservation needs and status of the bovids of Southeast Asia.

McShea, William J., Stuart J. Davies, and Naris Bhumpakphan, eds. *The Ecology and Conservation of Seasonally Dry Forests in Asia*. Washington, DC: Smithsonian Institution Scholarly Press and Rowman and Littlefield, 2011.
A fascinating account of the ecology of the dry forests, from their structure, dynamics, and floral composition to the elephants, wild cattle, deer, and tigers that call it home to the use and management of the forest by local communities and the role of fire in the ecosystem.

Sterling, Eleanor Jane, Martha Maud Hurley, and Le Duc Minh. *Vietnam: A Natural History*. New Haven, CT: Yale University Press, 2006.
The only text on the natural history of Vietnam, this book is easy to read for the nonspecialist. It highlights the amazing biodiversity of Vietnam and explains the conservation challenges that Vietnam faces.

TRAFFIC, 2008. "What's Driving the Wildlife Trade? A Report of Expert Opinion on Economic and Social Drivers of the Wildlife Trade and Trade Control Efforts in Cambodia, Indonesia, Lao PDR, and Vietnam." East Asia and Pacific Regional Sustainable Development Discussion Papers. Washington, DC: World Bank, East Asia and Pacific Region Sustainable Development Department, 2008.
The greatest threat to wildlife across the Indochina region is the rampant and largely uncontrolled wildlife trade. This important report documents the drivers of this trade, highlighting the scale and complexity of the issue from both the demand and enforcement angles.

Chapter 9. Rarity Made Common
Inskipp, Carol, Tim Inskipp, and Richard Grimmett. *Birds of Bhutan*. London: Christopher Helm, 1999.
The definitive guide to birds of this nation.

Seidensticker, John, Eric Dinerstein, Surendra P. Goyal, Bhim Gurung, Abishek Harihar, A. J. T. Johnsingh, Anil Manandhar, et al. "Tiger Range Collapse and Recovery at the Base of the Himalayas." In *The Biology and Conservation of Wild Felids*, edited by David W. Macdonald and Andrew J. Loveridge, 305–23. Oxford: Oxford University Press, 2010.
An account of rapid loss of tigers and recovery in a habitat that supports among the highest densities of tigers on Earth.

Wangchuk, Tashi, Phuntso Thinley, Karma Tshering, Chado Tshering, and Deki Yonten. *Field Guide to the Mammals of Bhutan*. Thimpu: Bhutan Trust Fund for Environmental Conservation, 2004.
The first comprehensive guide to the mammals of Bhutan.

Wikramanayake, Eric, Eric Dinerstein, John Seidensticker, Susan Lumpkin, Bivash Pandav, Mahendra Shrestha, Hemanta Mishra, et al. "A Landscape-Based Conservation Strategy to Double the Wild Tiger Population." *Conservation Letters* 4, no. 3 (2011): 219–27.
The scientific underpinning for the ambitious conservation goal of doubling the number of wild tigers by 2022, adopted by the tiger range countries in November 2010 at the International Forum on Tiger Conservation, St. Petersburg, Russia.

About the Author

Eric Dinerstein is Lead Scientist and Vice President of Conservation Science at World Wildlife Fund-US. Over the past forty years he has studied bears, rhinos, tigers, bats, and plants and many other creatures around the globe, and he remains active in the conservation of rare species. He has published over one hundred scientific papers and several books, including *The Return of the Unicorns: The Natural History and Conservation of the Greater One-Horned Rhinoceros* and *Tigerland and Other Unintended Destinations*. In 2007, *Tigerland* won the American Association for the Advancement of Science's award for science writing, the AAAS/Subaru SB&F Prize for Excellence in Science Books.

Index

Page numbers followed by "f" indicate maps and illustrations.

About Island Press

Since 1984, the nonprofit Island Press has been stimulating, shaping, and communicating the ideas that are essential for solving environmental problems worldwide. With more than 800 titles in print and some 40 new releases each year, we are the nation's leading publisher on environmental issues. We identify innovative thinkers and emerging trends in the environmental field. We work with world-renowned experts and authors to develop cross-disciplinary solutions to environmental challenges.

Island Press designs and implements coordinated book publication campaigns in order to communicate our critical messages in print, in person, and online using the latest technologies, programs, and the media. Our goal: to reach targeted audiences—scientists, policymakers, environmental advocates, the media, and concerned citizens—who can and will take action to protect the plants and animals that enrich our world, the ecosystems we need to survive, the water we drink, and the air we breathe.

Island Press gratefully acknowledges the support of its work by the Agua Fund, Inc., The Margaret A. Cargill Foundation, Betsy and Jesse Fink Foundation, The William and Flora Hewlett Foundation, The Kresge Foundation, The Forrest and Frances Lattner Foundation, The Andrew W. Mellon Foundation, The Curtis and Edith Munson Foundation, The Overbrook Foundation, The David and Lucile Packard Foundation, The Summit Foundation, Trust for Architectural Easements, The Winslow Foundation, and other generous donors.

The opinions expressed in this book are those of the author(s) and do not necessarily reflect the views of our donors.